THE ASTRONOMY QUARTERLY LIBRARY

1

Pachart

THE ASTRONOMY QUARTERLY LIBRARY

Editor

Eric R. Craine

Pachart Publishing House

THE GEMINI SYNDROME

Star Wars of the Oldest Kind

Roger B. Culver

Colorado State University

Philip A. Ianna

University of Virginia

Pachart Publishing House

TUCSON

Library of Congress Catalog Number: 78-059209
International Standard Book Number: 0-912918-17-9

Photograph Credits:

p16.   Reproduced by Courtesy of the Trustees of the British Museum.
p43.   NASA.
p47.   Hale Observatories.
p50.   Yerkes Observatory photograph.
p51.   Sproul and McCormick Observatories photographs.
p52.   The Kitt Peak National Observatory.
p95.   Bonnie Culver.
p96.   University of Texas McDonald Observatory.
p159.   Lowell Observatory photograph.

cover design: R.B. Culver
cover art: Jack Hadley

Pachart Publishing House
1130 San Lucas Circle
Tucson, Arizona 85704

"It's not what's in the stars that counts,
it's what's in the food dish."
--Morris the 9-Lives Cat

Preface

There are probably few astronomers who have not been mistaken at one time or another for astrologers. Occasionally people have asked us about their horoscopes, much as a doctor or lawyer is pumped for professional advice at a cocktail party, and then were visibly surprised when we admitted our somewhat limited skills in this area. Other sources of evidence indicate that in general the level of astrological interest in our society is quite high. Moreover, current interest in astrology among college students is sufficiently great that it has even been used as a successful vehicle for the teaching of *astronomical* principles.

It has also been our experience that there are many individuals who are most eager to hear objective scientific views regarding astrology. Unfortunately the attitude of the scientist is all too often one of detached disgust and most have refused to get involved in something which to them seems obvious nonsense. The danger in such an attitude is that it has tended to create a vacuum which the astrologers have easily filled in recent years. Even in the midst of the greatest resurgence in astrology since the scientific revolution, the response of science to the astrological phoenix has, to date, been largely disjointed and ineffective. Only two book-length works and a few scattered chapters in elementary astronomical texts currently offer any alternative to the plethora of astrological publications which unanimously sing astrology's praises. Even the much-publicized Humanist Magazine anti-astrology statement of September 1975 which was signed by 186 scientists, including 19 Nobel Prize winners, has had little effect on the resurgence of astrology. As indeed it should not have. We ideally should be persuaded only by hard evidence and not by dogmatic statements from on high. We note in passing that we didn't sign the statement. One of us (RBC) didn't even know about it until well after the fact. (Such are the hazards of teaching on the high prairie!)

Both of us have long been intrigued by the astrological boast that no one who has carefully studied astrology has failed to become convinced that it works. We started admittedly from a somewhat skeptical viewpoint, but nonetheless with a willingness to be shown there was some evidence favorable to astrology. After several years of consideration, including some data collecting and evaluation of our own, we are no longer merely skeptics.

We have no delusions that the contents of this book will provide a "knock-out" blow to astrology. Others with powers and abilities far greater than ours have tried and failed in this endeavor throughout astrology's 5000 year run on the planet. Astrology is clearly a firmly entrenched belief-system and its dedicated adherents are unlikely to be influenced by nonsupportive evidence. At least we hope to have made our presentation in such a way that the problems science currently has in accepting this durable craft are set forth for the benefit of astrologer and non-astrologer alike. In particular, we hope we have answered the many questions which have been asked of us down through the years regarding the present relationship between science and astrology. Within the constraints of time and library we have tried to examine original sources and to cite such references so that one need not take our word alone. The responsibility lies with the reader who sincerely wishes to know the truth — he must investigate for himself.

The writing of this book has been a most interesting experience, first of all because we have seen both scientists and astrologers operating at their finest levels as well as their shabbiest. We can smile, for example, when Joseph Goodavage in "Astrology: The Space Age Science" refers to Edwin Hubble as Carl Hubbell, if for no other reason than the fact that both men had a comparable number of 'career wins'. After all, we, ourselves in an article for the Astronomy Quarterly made what could easily be interpreted as a Freudian slip when we accidentally referred to astrologer Carl Jayne as Carl Payne!

The smiles come a little less readily, however, when we recall the attitudes of some of our supposedly rational and objective colleagues in the sciences. We have, for example, been criticized on a number of occasions for "wasting our time" in writing this book. One of us was verbally assaulted in a faculty meeting by a senior professor of history for even considering the discussion of astrology in a course to be given at "Mr. Jefferson's University". In another case known to us a psychiatrist interested in testing astrology as a diagnostic tool was strongly urged by his administrative superiors not to become involved.

There are other lessons to be learned from the astrological renaissance. During the course of our investigation we have found that far and away the astrologers' longest suit is their rapport with the general public. A few years ago, for example, astrologer Carroll Righter began the practice of referring to individuals born under the sign of Cancer as "moon-children", rather than "Cancers" owing to the latter's obvious distasteful connotations. And in what has to be a marvelous microcosm of the entire situation, we simultaneously ordered Sydney Omarr's "Astrology: You and Your Love Life" and Lawrence Jerome's "Astrology Disproved". Mr. Omarr's book arrived three days later and cost $1.50. Mr. Jerome's book arrived three weeks later at ten time this price.

There are, of course considerable problems in casting one's lot too much with the public's taste. The popular astrologers' stubborn adherence to the simplicity of sun-sign astrology is perhaps the premier example of the astrological willingness to barter scientific evidence for public acceptance. As scientists, of course, we do not have this freedom of motion since we are, of necessity, constrained by the results of our experiments. We are thus faced with the non-trivial task of popularizing such items as relativity and quantum theory. Unfortunately most scientists simply throw up their hands and exit the scene, thereby making their own little contributions to the late-night movie stereotype of the socially insensitive scientist as well as to the high ratings which science courses enjoy in the "top ten" list of most feared courses in the curriculum. The 19th century fissure between science and society has, in the 20th century, widened at an alarming rate. The astrological renaissance is yet another warning for us to start closing it up.

There are many persons without whose efforts this work would not have reached fruition, and so we would like to acknowledge their contributions. Thanks are due to those institutions and individuals who gave their kind permissions to use many of the illustrations and quotations used herein, and for his unfailing support in this endeavor, we wish to extend our gratitude to Dr. John Raich, Chairman of the Physics Department at Colorado State University. We would also like to extend our gratitude to our three typists, Sharon Perpignani-Huth, Ann Tipton, and Mary Spencer for converting our scribbled thoughts into readable manuscript, to the personnel of CSU's Educational Media department for their capable assistance with our illustrations, and to Dr. Eric Craine of Pachart Publishing House for his editorial assistance. Other individuals who have contributed advice, comments, or material include (Mrs.) Johnny Demetsky, Marty Altschul, Rick Babarsky, Barbara Sanford (wherever you are), Ann Guinan, Pat Hammond, Kerry Kingham, Al H. Morrison and Charlie Tolbert. Lastly, we wish to thank astrology itself for providing us with the five millennia of sidereal lore which has made this book so enjoyable to write.

Roger B. Culver, Ft. Collins, Colorado
Philip A. Ianna, Charlottesville, Virginia

TABLE of CONTENTS

Figure 1.1   Some of the current astrological paraphernalia.

Chapter 1

The Gemini Syndrome

The heavens hold a virtually limitless treasure-trove of fascinating objects. The sun, the moon, the slowly moving bright planets, and the thousands of flickering stars all have intrigued human beings for thousands of years. As with any treasure of vast riches many have sought the key to the hidden wealth. Down through the ages, the seekers of understanding have chosen to follow two separate paths and have imagined the search for two very different sorts of reward. Thus, the human captivation with the heavens has produced the fraternal twins of astronomy and astrology.

In spite of some public confusion over the two endeavors, these twins are not identical by any means. Astronomy and its major subfield, astrophysics attempt to describe and understand the formation, behavior and "fate" of the planets, stars, and galaxies that comprise the observable universe. It is the world's oldest science; it has also been described as the world's second oldest profession. Astrology is as ancient but is in principle different. It is concerned with the behavior and fate of people, and sometimes their institutions, in a rather special way. Other than on a social level the *only* point of overlap is in the mutual use of the positions of celestial objects on the sky. The noted astrologer (and composer) Dane Rudhyar has defined astrology as the "study of the correlations that can be established between the positions of celestial bodies around the earth and physical events or psychological and social changes of consciousness in man" (1.1). He has elsewhere described astrology as "a technique of symbolization and of prognostication" (1.2), and the study of "an observable parallelism between the *timing of events* (author's italics) in the universe and in the individual consciousness" (1.3). To add further clarification Rudhyar says "astrology does not attempt to tell in a scientific manner how things happen" (1.4), that is, it has no interest in the physical causes and effects in celestial phenomena, but

1

rather deals with "their interpretation in terms of human character and behavior" (1.5). This seems to us to adequately represent the prevailing view. However, a fundamentally different definition has been given by Nicholas de Vore, a past president of the Astrological Research Society. According to Mr. de Vore, astrology is "the science which treats of the influence upon human character of cosmic forces emanating from celestial bodies" (1.6). The correlational and "cosmic forces" definitions both appear in current astrological literature and each has its supporters. We shall subsequently concern outselves with both views.

It has disturbed some astronomers to observe a resurgence in the public's interest in and awareness of astrology. The Gallup Poll for Oct. 19, 1975 places 32 million Americans in the category of those who "believe that the stars influence people's lives" (1.7). That same poll indicates that 77 percent of all Americans and 90 percent of all Americans under the age of 30 years can identify the astrological "sun-sign" of their birth. A recent poll of students at the Colorado State University yielded similar results. By contrast, that same poll indicated that less than 60 percent of the students knew their blood type and less than half their normal blood pressure! According to a recent tabulation (1.8) there are over 10,000 practicing astrologers in the United States with perhaps 1000 full-time professionals, 20 astrological jounals, and more than 1000 newspapers carrying astrological forcasts (1.9). There are three national organizations in the USA and over 150 local associations. It has been further estimated that five million Americans spend 200 million dollars each year consulting astrologers. Similar figures can be found for other countries. By contrast there are several thousand members of the American Astronomical Society and, omitting the space program, perhaps 100 million dollars a year is spent on research in astronomy. There is virtually no bookstore, newsstand, or public library which does not carry an array of texts, booklets, and magazines on the subject of astrology. Nor is there any aspect of life for which astrological advice is not at the ready. There is astrological guidance regarding careers, finances, health, and psychological problems. One can even get signs from the stars on sex, gambling, and the choice of an appropriate perfume (1.10).

A number of words in the English language such as *disaster, consider, aspect, martial,* and *lunatic* come to us from the astrologers. In recent years, one of the pieces of information listed along with the Miss USA contestants' more traditional "vital" statistics is the sun-sign of a given contestant. While drinking Zodiac Malt Liquor*, one can also enjoy "Astrology – The Game of Prediction"** (which, naturally, is played with a zodiacal die!). A recent novel entitled *The Deadly Messiah* by David and Albert Hill has as its heroine a brilliant astrologer whose astrological skills and prowess far outstrip those of her contemporaries in the real world. In another example, one minute horoscopes are just a telephone call away thanks to "Jeanne Dixon's Horoscope-By-Phone." In short, we are in the midst of an astrological renaissance such as the world has not seen in over four centuries.

*Peter Hand Brewing Company, Chicago, Illinois 60622.
**Copyright 1972, Dynamic Design Industries, Anaheim, California 92803.

As astrology has grown, there has been a systematic attempt in the astrological literature to downgrade the opinions of not only astronomers but scholars of other fields as well who are deemed to be "anti-astrology." Some examples:

" ... physicists continue to be cynical or antagonistic to astrology, even when their own world was shown to be other than they thought ... they have built up a lengthy record of error, but persist in refusing to accept the occult ... " (1.11)

" ... an astronomer knows no more about astrology than a radio mechanic knows about music. To ask an astronomer for his 'expert' opinion on the subject is useless." (1.12)

"Like many of his colleagues, Dr. Marshall (then director of Fels Planetarium) seemed actually to believe the distortions in some of our current textbooks on the history of science — distortion originally encouraged by astronomers to cover up the fact that their most illustrious predecessors were astrologers ... " (1.13)

"In spite of all the scientific abuse that has been heaped upon astrology, the world today is manifesting an active interest in it ... On the record, it has not been the 'crack-pots' who have testified to astrological convictions but the acknowledged great; and its critics, for the most part, are to be found only among the near-great." (1.14)

"It is as much the responsibility of scientists to prove their negative statements concerning Astrology as their positive statements in their respective fields. To live in a world of smug assumptions of one's own superiority and refuse to see any need for hard investigations of the phenomena outside their domain, under the false dignity of 'science', only encourage the unholy development of the Cult of Science." (1.15)

The scientific response has been limited for the reason that most scientists have considered astrology to be so clearly false as not to be worth the effort of investigation. For example in *Objections to Astrology,* Dr. Bart J. Bok comments:

"At one time, I thought seriously of becoming personally involved in statistical tests of astrological predictions, but I abandoned this plan as a waste of time unless someone could first show me that there was some sort of physical foundation for astrology." (1.16)

No amount of invective from either side is of any benefit to those individuals who honestly seek information on astrology and its principles. However, with few exceptions they can presently only turn to the vast quantity of information that emerges from astrological sources. The extensive array of popular astrological texts is balanced by only three similar works which discuss to some degree the scientific position on astrology: *Objections to Astrology* by Bok and Jerome is essentially a reprint of a discussion first appearing in the September/October 1975 issue of "The Humanist"; Delano's *Astrology: Fact or Fiction* (1.17), includes the scientific objections to astrology as part of a general

3

assault on the topic; and most recently published, Jerome's *Astrology Disproved* (1.18). Other works such as Eisler's *The Royal Art of Astrology* (1.19) tend to be somewhat out of date in light of recent developments in the debate over astrology.

Any truly adequate discussion of astrology must necessarily be based on a detailed knowledge of astrology and its history, principles, methods, and techniques, but most importantly it must be based on the available empirical evidence. Because of the dearth of information available on astrology in the scientific literature we have had to collect results from a wide variety of sources. These range from the Sunday Supplements and the "National Enquirer" at one end of the spectrum to technical journals when possible. We hope we have taken a step toward the proper evalutation of astrology that will satisfactorily demonstrate to many whether or not astrology works.

The craft of astrology, like astronomy, can be learned at various levels. For the advanced practitioner the subject can be quite complex and require a firm background in basic mathematics such as spherical trigonometry. Years of study may be invested before the discipline is fully mastered. There are a number of astrological institutes and groups offering academic facilities and courses for the serious student of astrology. The Faculty of Astrological Studies in London requires its students to pass a series of examinations (1.20). You may earn a Certificate after one year of study and a Diploma after two. Those who have received the Diploma are entitled to put the abbreviation "D.F. Astrol. S." after their names. Another certifying agency is the American Federation of Astrologers. Both also have ethical codes, sort of astrological Hippocratic oaths to which their students must subscribe, providing guidelines for their professional conduct. So there are some similarities with the educational process in astronomy where the certification for research generally involves obtaining a Ph.D. from some suitable institution, and usually takes four or five years.

Some of the more popular astrologers may have a somewhat less formal educational background. Linda Goodman admits her knowledge is based at least in part on her having been an astrologer in a previous reincarnation (1.21), and the Great Zolar was a men's wear salesman who learned to cast horoscopes in Atlantic City, New Jersey from a boardwalk astrologer (1.22). It should be emphasized that this is not typical of the most serious astrologers who have followed the more arduous path of rigorous schooling.

Like any enduring body of knowledge, astrology has its own set of basic definitions and concepts. These may not be familiar to every reader so it would perhaps be useful to introduce some of them here. We shall devote two of the following chapters to similar discussions of 'science' and analogous astronomical concepts. But astrology first.

All of the fundamental elements of interest are in one way or another related to the heavens which surround us like a great crystal sphere. The apparent annual path of the sun around the celestial sphere (not its daily path) is called the ecliptic (See Figure 1.2). It is a circle of 360 degrees centered on the earth, and it prescribes the modern shape of the horoscope. In addition to the sun, the moon and all but one of the planets can be found within the *zodiac,* an 18 degree wide belt of sky centered about the ecliptic. Owing to the 17 degree tilt

4

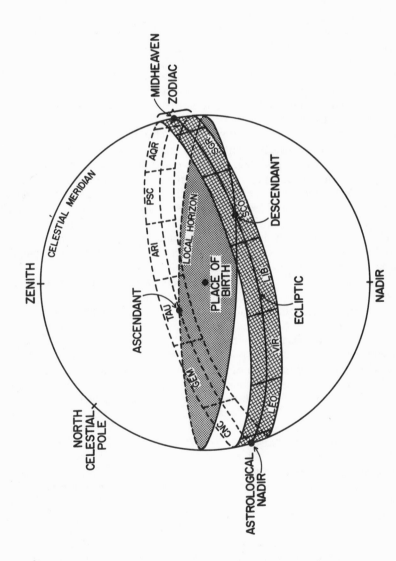

Figure 1.2   The geometry of the astrological reference points and circles for a latitude of about 41°N.

5

of its orbit, Pluto is the exception, spending the bulk of its time outside of this band. The astrological ecliptic is divided into twelve equal 30 degree arclengths, the houses of the planets, also known as the *signs of the zodiac*. They are subdivisions of the sun's apparent annual motion about the earth, the sun occupying a different sign for one month at a time. The signs start at the vernal equinox, the sun's location at the beginning of spring, or 0 degrees Aries. In order toward the east, the 12 signs include Aries, Taurus, Gemini, Cancer, Leo, Libra, Virgo, Scorpio, Sagittarius, Capricorn, Aquarius, and Pisces. There is some disagreement among the astrologers over the location of the starting point of the astrological signs relative to the astronomical constellations bearing the same names, because the vernal equinox does not remain fixed relative to the stars (see Chapter 6). Historically, this starting point seems to have been set on the eastern border of the sign of Aries, or the so-called "first point of Aries," although some have argued for an original zodiac starting in Taurus. The *tropical* astrologer positions this starting point so that it is always coincident with the vernal equinox point regardless of the stars and constellations that may form the background. The *sidereal* astrologer, on the other hand, regards the astrological signs as being roughly coincident with their constellation namesakes.

The signs are subclassified into three families according to certain assigned qualities. Divided alternately into positive and negative (or active-passive) signs they first form *polarities*. As usual we start with Aries (positive), then Taurus (negative), and continue around the zodiac. The modes, or *quadruplicities* are three groups of *cardinal, fixed,* and *mutable* signs, again starting with Aries and labeled clockwise around the zodiac. The four ancient elements, fire, air, water and earth, apply to the third family of elemental signs, or the *triplicities*. They are identified in a manner similar to the previous division. For example, the fire signs are Aries, Leo, and Sagittarius.

We may also overlay the astrological zodiac with a second subdivision into 12 parts numbered sequentially counterclockwise from the point analogous to the east point on the horizon. These divisions of the daily cycle of rotation of the celestial sphere are called the astrological *houses*, each representing a certain field of life experience. To complicate matters this division can be accomplished in a multitude of ways. They also may be grouped into the same three families as the signs. We shall discuss the various house systems in more detail later.

The signs and houses provide the basic background frame of reference with which to evaluate the celestial 'correlations' or 'influences'. Against this background a map is drawn of the relative locations of the sun, moon, and planets in the sky for a specified moment and place. This plot is of course the horoscope. It may be "cast," or "erected" for the moment of birth, or for some other time, depending on the purpose. For a moment-of-birth or *natal* chart, the time must be accurate to the nearest minute and the latitude and longitude of the birthplace must be known. For example, in Figure 1.3, the birth chart for September 6, 1940 is given with the planets located in the signs and houses.

The astrologer attaches special significance to a number of features of the horoscope. The signs and houses in which the sun and moon are located are of great importance, as well as the placement of the remaining planets. Sometimes the nodes, that is the intersections of the moon's orbit and the ecliptic, may be

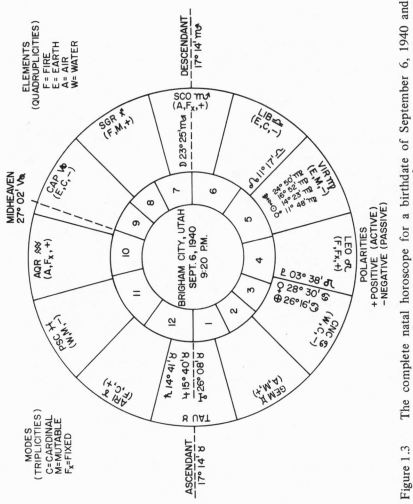

Figure 1.3   The complete natal horoscope for a birthdate of September 6, 1940 and a birthplace at 41°31' North latitude and 112°01' West Longitude. The modes, elements and polarities are indicated for each sign.

7

included. The influence of a given body will be different if it is close to a *cusp*, or dividing line between one house and another.

The *angles* are especially sensitive spots in the chart. These are the four cardinal points as plotted on the horoscope: the east point, or *Ascendant*; the west point, or *Descendant*; the mid-heaven, or *Medium Coeli* (or M.C.) variously defined as the south point, or the point where the sun reaches its highest point at noon, or the cusp of the tenth house; opposite to the M.C. is the *Imum Coeli* (or I.C.), the north angle, or cusp of the fourth house sometimes. The most powerful angle is generally considered to be the Ascendant.

Also of considerable consequence are the *aspects*, the angular relationships between the sun, moon, and planets, as measured by the separation between any two objects along the ecliptic. They will be either *applying*, (or forming) if the spacing between the objects is decreasing, or *separating*. The major aspects and their angular separations are the conjunction ($0°$), semi-sextile ($30°$), semi-square ($45°$), sextile ($60°$), square ($90°$), trine ($120°$), sesquiquadrate ($135°$), quincunx ($150°$), and opposition ($180°$). Additional aspects that may occasionally be noted are those with separations of $7½°$, $15°$, $22½°$, $40°$, $51½°$, $102½°$, $144°$, and $154°$. Since the actual angular spacings seldom occur exactly as defined, astrologers consider an aspect to be effective over a certain range. Accordingly each planet has an *orb* associated with it. These may be as small as $5°$ for Neptune and as large as $17°$ for the sun. Different astrologers allow different values for the orbs. To find the range of effectiveness of two planets the orbs are combined. When approaching conjunction for example, the respective orbs are first added together and then divided by two. The interpretation of these interactions depends on whether the aspect is considered to be strong or weak, malefic, or benefic, or if other special circumstances exist, such as a planet being with about $5°$ of the sun, or *combust*.

The final determination of the potentialities in a horoscope is ideally based on a *synthesis* of all the many factors exhibited in the chart. The distribution of the planets is examined for groupings and balance between the right-left and upper-lower hemispheres. A count is made of the number of planets occupying the signs and houses separately dividing them first by active-passive signs and houses, then by the quadruplicities and triplicities. A planet's influence may be strengthened or weakened by its placement in the chart. Such conditions are termed *dignities* or *debilities*. There are at least several dozen of these. A planet falling in the sign of which it is considered to be the *Ruler*, such as Mars in Aries, is strengthened; if it falls in the sign opposite to the one of which it is said to be Ruler, that is in its *Detrement*, it is weakened. Special attention is paid to the Ascending sign. The nature of each planet, including the sun and moon, its influence according to the sign and house in which it appears, its location within the sign or house (is it near a cusp or any of the angles?), and how it is aspected by other planets must be noted in detail. Although this list is not exhaustive, it is possible to appreciate how the final blending of all the elements in the chart may require many hours of an experienced astrologer's time.

The horoscope and its interpretation most frequently involve an individual and his life. The daily newspaper horoscope is a widely seen but inferior example of this. However, astrology is applied much more broadly and consequently it

8

can be regarded as having a number of branches. *Judicial* astrology is the general division that interprets celestial motions as affecting terrestrial life. Included here is *natal* or *genethliacal* astrology which deals with the casting and interpreting of individual birth charts. *Mundane* astrology has application to large groups of people and answers questions concerning cities, states, and national destinies. If you need a specific answer to a specific question regarding some situation or event, perhaps a business matter, some personal affair, or even a particular thought, you would turn to *horary* astrology. This branch of astrology has its own set of rules and guidelines. The horoscope is cast generally for the birth moment of the idea or question. A related field is *electional* astrology. It assists you in selecting a propitious moment for the start of some new endeavor or project, for example when to get married, or take a trip, or when to begin a new business or build a building. In *medical* astrology the horoscope becomes a diagnostic tool to aid in the evaluation of a person's health and identify illnesses. Finally *meteorological* and *agricultural* astrology have self-evident definitions.

The prediction of events and effects resulting from the passage over certain points in the chart of the various moving bodies of the solar system is a major part of astrology, especially in the genethliacal and mundane forms. There are a wide variety of methods used by astrologers to accomplish such prognostication. The three basic ones seem to be those of transits, primary directions, and secondary progressions. The system of *transits* is based on the actual motions of the planets. As the planets move at their normal (predictable) rates around the sky, certain consequences are expected at any time a given body passes over a sensitive point in the chart, such as for a given day Jupiter transiting the original birth position of Venus. What might be expected to transpire is of course dependent upon the object, the point of passage, and the interpretation by the astrologer. On the other hand directions and progressions foretell events by analogy. In *primary directions* it is assumed that a degree of arc along the ecliptic is equal to a year. That is if Jupiter is placed 15° from Saturn in the birth chart, and if Jupiter is "directed" to the position of Saturn, an event of some import is expected in the 15th year of the person's life. Such "directions" are applied to each planet, the angles, etc. in turn yielding a large number of predictions. The *progressed* horoscope treats the planet's advance in the chart as if one day of motion were one year of the individual's life. So in the natal chart when you find the Moon conjunct Mars 42½ days after the moment of birth this represents the possibility of some significant happening in the life of this individual at age 42½ years.

Such planetary manipulations also provide a means of arriving at the correct moment of birth in those cases for which an accurate time (to one minute) is not available, or when the given birth data may require verification. This process is called *rectification*. The birth moment is checked by noting some especially signficant happening in the individual's life such as a serious accident or other event and its time of occurence. A progressed chart feature or transit representative of this event is found and used to alter or define the moment of birth so the real life and horoscope events coincide. There are also other methods for rectifying the birth chart. For example one other procedure

9

involves examining the houses in which the planets fall and comparing them with the person's circumstances and so noting what adjustment of the birth time would give better agreement in the chart.

This short description hardly does justice to the rich and complex 'treasures' of horoscope interpretation. Indeed, Noel Tyl's twelve volume series *The Principles and Practice of Astrology* remains introductory, and many authorities must be consulted before anything nearing a complete picture emerges. A contributory factor here is the addition of new methods and features over the thousands of years of history by each new generation of astrologers. It is not that astrology has been refined nor that its capacity for illuminating the human predicament has improved. Rather, in its quest to achieve its ends, astrology has groped more widely for the key to understanding while casting aside nothing. Indeed there is much here in common with religion. The faithful have added new procedures and dogma, seldom if ever doubting or questioning the fundamental beliefs. Astronomy, too, has experienced an enormous increase in its complexity, but with a basic difference. Astronomical studies of the heavens have led us to a more detailed and exact knowledge of the universe around us, but they have also forced the discarding of a great many myths and mistakes in the course of the journey. Thus, crystalline spheres, a hollow sun, and comets as harbingers of catastrophe have all been cast aside by scientists, while the combust, semi-sextile, and part-of-fortune are still cherished and revered as astrological tools the world over. In these and many other ways the siblings astronomy and astrology are opposites. We expect in what follows to show that in personality, astrology is resistant to change, has vague but usually well-intentioned ideas, is too easily susceptible to outside influences, is often preoccupied with the pursuit of money, is quickly aroused, frequently irrational, and almost always wrong.

1.1   Rudhyar, D., "The Astrological Houses", p. 3, Doubleday and Company, Inc., Garden City, NY (1972).

1.2   Rudhyar, D'., "The Practice of Astrology", p. 10, Penguin Books, Inc., Baltimore, MD (1968).

1.3   Ibid., p. 11.

1.4   Ibid., p. 11.

1.5   Ibid., p. 5.

1.6   de Vore, N., "Encyclopedia of Astrology", p. 28, Littlefield, Adams, and Co., Totowa, NJ (1976).

1.7   Bok, B.J., *Physics Today*, Jan. 1977, p. 84.

1.8   Dean, G. and Mather, A., "Recent Advances in Natal Astrology", p. 6, The Astrological Association, Bromley Kent, England (1977).

1.9   *Time*, Dec. 28, 1962, p. 47.

1.10   Carter, M., *Mademoiselle* **82**, 34 (1976).

1.11   Stone, H., *Astrology: Your Daily Horoscope*, Mar. 1977, p. 40.

1.12   West, J. and Toonder, J., "The Case for Astrology", p. 127, Penguin Books, Inc., Baltimore, MD (1970).

1.13   Omarr, S., "My World of Astrology", p. 95, Wilshire Book Co., (1975).

1.14   Ibid., Ref. 1.6, p. viii.

1.15   Bhujle, V. V., CAO Times, **3**, 23, 1978.

1.16   Bok, B. and Jerome, L., "Objections to Astrology", p. 31, Prometheus Books, Buffalo, NY (1975).

1.17   Delano, K., "Astrology: Fact or Fiction", Our Sunday Visitor, Inc., Huntington, IN (1973).

1.18   Jerome, L., "Astrology Disproved", Prometheus Books, Buffalo, NY (1977).

1.19   Eisler, R., "The Royal Art of Astrology", Herbert Joseph Publishers, London (1946).

1.20   MacNiece, L., "Astrology", p. 228, Doubleday and Co., Inc., Garden City, NY (1964).

1.21   Weisinger, M., Parade, June 3, 1973, p. 5.

1.22   Ibid., Ref. 1.18, p. 5.

Chapter 2

Star Wars of the Oldest Kind

"Astrology unquestionably arose, in the beginning, as a wholly empirical science. Man began to observe correspondences between the events in his life and the seasons—or the phases of the moon, and other celestial phenomena—and to organize the correlations to the extent of his intellectual power" (2.1). The claim that astrology originated as an observational science with earliest man is one of the commonly recurring themes in astrological literature. The astrologers believe, and would like us to believe, their procedures emerged from some foundation in observational fact. It would be quite reassuring if it were so, but this viewpoint has little support. In the beginning, astrology was neither "wholly empirical" nor "science"; it arose from a blend of imagination, fear, and religious superstition.

A second theme frequenting astrological writing is that astronomy came second in history as "the haughty scientific offspring of astrology" (2.2). The historical record is of course quite incomplete, but it is clear that one could argue with equal verity that astrology is the arrogant religious offspring of astronomy. There is, in fact, a certain amount of evidence suggesting an earlier origin to astronomy proper and a greater degree of independence between the two than is sometimes thought. The astrologers also seem to derive a certain amount of satisfaction, gleefully at times, from pointing out the numerous historically important astronomers who, it seemed, also practiced astrology. Although from some there later came rejections, considered in the context of their times, their involvement with astrology is of little consequence. Science after all developed as a trial-and-error procedure, and often one small step of true progress has followed only after many giant false ones.

The civilization that brought us astrology flourished in Mesopotamia, the land between the rivers, more than 4000 years ago. Its roots extend back thousands of years more into Near East antiquity to the beginning of architecture and astronomy, of written history and mathematics. From our distant vantage point we look back at the history of this area and it appears like a beach swept over time and again by the tide of each newly rising city state and invaders from beyond the valley of Mesopotamia. There were continuing battles for territory and dominance, and the arts and sciences developed it seems in spite of apparent chaos.

Five thousand years ago, near the Euphrates in southern Babylonia, the Sumerian civilization climbed toward its zenith within such walled cities as Ur, Larsa, Erech, and Uruk. About this time cuneiform writing was invented and a method of writing numbers was developed in response to the needs of commerce. The Sumerians had multiplication tables, tables of squares, square roots, cube roots, and reciprocals. There was a calendar based on the lunar month, not quite a year in length. Each city built elaborate brick temples to its gods of sun, moon, earth, and fertility, and gave its own names to the months. To the Mesopotamian of that day, each event and object around him had an independent will and personality of its own. The fate of all beings and things was decided by an assembly of gods who were the controlling forces of nature. Their leader was Anu, god of the heaven, and next to Anu was Enlil, god of the storm.

In central Mesopotamia, around the city of Agade, formerly nomadic tribes settled and prospered through trade. About 2350 B.C., Sargon I, having brought the several city states of Akkad and Sumer together, moved into the north. Uniting Mesopotamia and adding some of the surrounding area he thus formed one of the first great empires of history. The Akkadians adopted much from the Sumerian culture. They adapted cuneiform to the writing of their Semitic language, and they took up the mathematical and business methods of the Sumerians. They worshipped the sun, the moon, and the planets, especially Venus. The dynasty of Sargon declined and in about 2200 B.C. the barbarian Guti tribe overran Akkad. They in turn were expelled from Mesopotamia by Sumerians from the city of Erech, and shortly thereafter kings of the third Ur dynasty re-established the area claimed by Sargon. However invasions by Elamites from the east and Amorites from the west left Mesopotamia again shattered into independent city states.

From Babylon the Amorite King Hammurabi conquered the surrounding countryside to form the third extensive Mesopotamian empire, one of unprecedented power. Hammurabi reigned from 1728 B.C. to 1686 B.C. During this period he imposed a code of justice in part derived from older Sumerian laws. The land was irrigated and Babylon became a center of commerce and culture. Medicine, mathematics, and physical sciences advanced, and there was literature and music; paintings decorated walls and temples. But this too was not to last. Under Hammurabi's successor and son, Samsu-Iluna, the power of Babylon declined. Finally Babylon was invaded and sacked by the Hittite king Mursili I in 1530 B.C. Shortly thereafter the Kassites moved in from the east and held power in the central third of the region until about 1160 B.C. The sciences progressed little during this unsettled period.

Still farther to the north Assyria grew in strength. The Warrior Assyrians

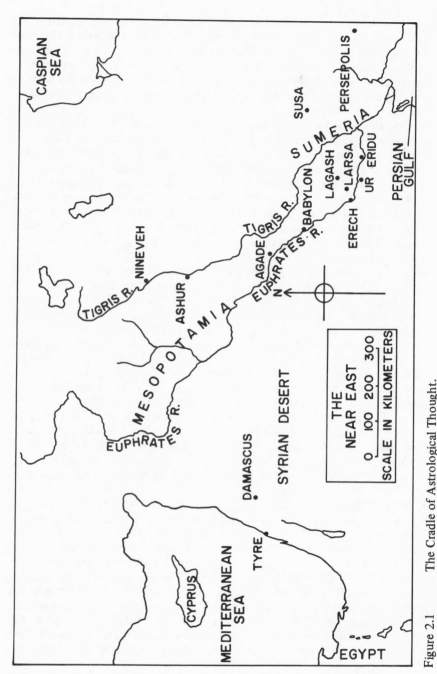

Figure 2.1   The Cradle of Astrological Thought.

14

struggled continuously and often cruelly in series of wars against the surrounding states. By 800 B.C. they were the most powerful state of the region. As had happened with the Akkadians, the warlike Assyrians adopted much of the culture of Babylon. And as their wealth from conquest grew, so did the interest in the arts and sciences. Ashurbanipal (668-630 B.C.) was as keen a warrior as the kings that preceded him, but he brought literature and learning to a zenith during his reign as well. His scholars and scribes combed the country collecting and copying the ancient tablets of the Sumerians and Akkadians. He built a great library at Nineveh containing tens of thousands of cuneiform tablets without which our knowledge of the religion and science of the time, especially astrology and astronomy, would be very meager. With the decline of the Assyrian kingdom, the Babylonians and Medes in 614 B.C. conquered the city of Assur, and in 612 B.C. they destroyed Nineveh. Babylon arose again under Nabopolassar and Nebuchadnessar (about 600 B.C.) until the Persians gained domination around 540 B.C. Under Alexander the Great, who conquered the Persians in 331 B.C., the Babylonian temples and priests were very active, and astrology flourished as never before.

Prior to recorded history, the interest of man in the sky was no doubt a very practical one. It must have involved daily time keeping, the development of a calendar, and knowledge of the changing seasons. Primitive hunters following migrating animals must have gauged their movements from celestial signs of the seasons and used the stars for guiding their own travels. Newly emerging agricultural societies depended on such knowledge to anticipate the times of floodings and planting. And so the science of observing the sky, astronomy, became the first science (2.3). We emphasize that this is astronomy. Astrology followed much later as the religions of Mesopotamia grew, embraced astral gods, and were extended to forecasting the fate of man.

The oldest known astronomical and astrological texts date from the first half of the Hammurabi dynasty. The astrological texts are simply omens based on the appearance of the moon and planets. An example from the earliest part of this period renders some general predictions related to the moon:

If the sky is bright when the New Moon appears and it is greeted with shouts of joy, the year will be good.

If on the day of the new crescent, the Moon-god does not disappear promptly from the sky, 'shivering' will come upon the land. (2.4)

Although these are among the oldest texts that we have, omen astrology is surely older than this. Omens which mention King Sargon I of Akkad and other rulers of that period have been preserved from the time of Hammurabi (2.5), probably as copies of much older writings.

The most extensive collection of astronomical observations and celestial-meteorological omens known, about 7000 of them, is contained in the 70 or so cuneiform tablets of the "Enuma Anu Enlil" series. There is evidence that they were collected sometime between 1350 - 1100 B.C. for the purpose of summarizing the astrological knowledge of the time (2.6). Preserved in the

Figure 2.2   Two Babylonian Horoscopes (Bottom pair) The horoscope for an individual who was conceived on March 17, 258 B.C. and born on December 15, 258 B.C. (Top pair) The horoscope for a person who was "born under the brilliant house of Jupiter" on March 1, 142 B.C. (Bottom pair)

16

library of Ashurbanipal, they had probably been in use for hundreds of years. These omens are considerably more sophisticated than those quoted above:

"If in the month Shabatu, the 15th day, Venus disappeared, and stayed out of the sky for three days, and became visible in the east on the 18th day of Shabatu, springs will open, Adad will bring his rain, Ea will bring his floods, and king will send messages of reconcilliation to king. (2.7)

It is clear from other omens in the Enuma Anu Enlil series that there is a fundamental tie between Babylonian religion and astrology. The planet Jupiter is referred to as "star god Marduk", and similarly Venus in half its dual role was Ishtar, goddess of love. Mars was Nergal, god of war. Van der Waerden has pointed out (2.8) that in the astrological omens, the movement of Mars is most frequently connected with war and destruction, and many Venus omens are associated with love and fertility. We give two examples:

When Mars approaches the star Shu.gi there will be uprising in Amurru and hostility; one will kill another . . .

When Venus stands high, there will be pleasure of copulation . . . (2.8)

But such omens must be placed more precisely in the context of the time, for celestial phenomena hardly commanded the undivided attention of the Babylonians. The astrological omens are but a fraction of the omen collection in Assurbanipal's library. To the Babylonians and the Sumerians before them the world and its wonders were menacing and magical. They saw portents and indications of the future in every event around them, the behavior of birds, insects, and animals, dreams, the entrails of animals, flour or oil dropped into a jar of water. They wore amulets and charms to frighten away demons. Incantations and therapeutic remedies abound for coping with many specific problems, including those sexual in nature:

"If a man becomes impotent in the month of Nisannu, you catch a male partridge, pluck its wings, strangle it, flatten it, scatter salt on it; you dry it, crush it together with seeds of the mountain-dadanu plant; you give it to him to drink in beer; that man will regain potency." (2.9)

In the Shumma Izbu omen series (2.10) alone there are about 2000 omens regarding unusual births. Keep in mind that the astrologers would have us believe astrology arose as the result of the observation of planetary patterns in the sky coinciding with events on the earth (2.11) as you read the following typical omens:

"If a woman gives birth to a pig — a woman will seize the throne." (2.12)

"If a woman gives birth to an elephant — the land will be laid to waste." (2.13)

17

"If a woman gives birth to two girls and they have only one abdomen — (there will be) dissension between man and wife." (2.14)

"If a ewe gives birth to a lion and it has two horns on the left — an enemy will take your fortress." (2.15)

"If I throw oil on water, the oil sinks and rises up and its water still covers it — in regards to the campaign on uprush of evil — for the sick person this means a divine hand has afflicted him." (2.16)

"If a man goes on an errand and a falcon passes from his right to his left — he will achieve his goal." (2.17)

"If a snake falls on a man — (there will be) attainment of desire." (2.18)

"If a white dog pisses on anyone, poverty will overtake him; if a black dog does the same, sickness will seize the man; if a brown dog does the same, that man will be joyful." (2.19)

There is obviously no fundamental difference between these omens and the celestial ones. They can surely have no empirical basis. For one thing, other than the biological difficulties, in spite of the existence of thousands upon thousands of tablets preserving both astronomical observations as well as the most routine business transactions, there are no known observational records from which omens like those above might have been derived. They appear to be a simple record of religious superstition, and they developed right along with the astrological omens as part of a consistent magical view of the world. Indeed Will Durant sums up nicely:

"Never was a civilization richer in superstitions. Every turn of chance from the anomalies of birth to the varieties of death received a popular sometimes an official and sacerdotal, interpretation in magical or supernatural terms . . .

The supersitions of Babylonia seem ridiculous to us, because they differ superficially from our own. There is hardly an absurdity of the past that cannot be found flourishing somewhere in the present." (2.20)

As mentioned above, following the Assyrians, the Babylonian kings Nabopolassar and Nebuchadnezzar II sought to restore the glory of Babylon. During this period (about 612 - 539 B.C.) the zodiac was divided into its 12 signs of 30 degrees each and systematic observations of the moon and planets were carried out. With the development of mathematical astronomy in the Persian period (539 - 331 B.C.), including the determination of the periods and the calculation of the motions of the moon and planets, the necessary theoretical framework existed for the first time to allow the construction of horoscopes

(2.21). Put another way, the tools with which to make astrological predictions that could then be observationally confirmed were not previously available. Again, the indications are that astrology was nothing like empirical science.

The oldest known horoscope comes from this period, and has been dated as April 29, 410 B.C. (2.22). It is definitely a genethliacal chart although only approximate planetary figures appear:

1. Month (?) Nisan (?) night (?) of (?) the (?) 14th (?)
2. son of Shuma-usur, son of Shuma-iddina, descendant of Deke was born.
3. At that time the Moon was below the "Horn" of the Scorpion,
4. Jupiter in Pisces, Venus
5. in Taurus, Saturn in Cancer
6. Mars in Gemini. Mercury, which had set (for the last time), was (still) in(visible).

The question marks indicate uncertainty in the translation or required words which have been inserted; the numbers are line numbers for the tablet. No interpretation appears with the list except that line 9 reads "(Things?) will (?) be good before you". Another genethliacal horoscope from 263 B.C. is similar but it includes a forecast for the child's life reading in part:

" . . . He will be lacking wealth. His food (?) will not (suffice?) for (his) hunger (?). The wealth he had in his youth (?) will not (remain?). The 36th year (or: 36 years) he will have wealth (His) days will be long (in number) . . ." (2.23)

There are other horoscopes for 258 B.C., 234 B.C., and later, so we have at least some idea when these astrological techniques developed in Babylonia and how they paralleled the development of astronomy. It is interesting that no Babylonian horoscope mentions the ascendant, or any other secondary astrological indicator (2.24).

Although originally "Chaldeans" were just those people living in the southern region of Babylonia, the term seems gradually to have been associated with a class of priest-diviners only one skill of which was astrology. They also foretold events by observing flights of birds, the entrails of animals, and interpreted dreams (2.25). Such were the times. Again, though the astrologers often point to the assumed common origin of astrology and astronomy, by no means, even by this time, were astronomers and astrologers necessarily the same person. The Greek historian and geographer Strabo (64 B.C. - 19 A.D.) wrote that:

"In Babylon, a settlement is set apart for the local philosophers, the Chaldaeans, as they are called, who are concerned mostly with astronomy; but some of them who are not approved of by the others, profess to be genethliaologists". (Strabo, Geography, (2.26) SVI, I).

Marcus Tullius Cicero writing in about 45 B.C. tells us of others. Eudoxos, a pupil of Plato who lived about 370 B.C. contributed much to mathematics and astronomy in early Greece. Cicero records that:

19

"... Eudoxus, whom the best scholars consider easily the first in astronomy, has left the following opinion in writing: 'No reliance whatever is to be placed in Chaldean astrologers when they profess to forecast a man's future based on the day of his birth.' Panaetius (about 140 B.C.), too, who was the only one of the Stoics to reject the prophecies of astrologers, mentions Anchialus and Cassander as the greatest astronomers of his day and states that they did not employ their art as a means of divining, though they were eminent in all other branches of astronomy. Scylax of Halicarnassus, an intimate friend of Panaetius, and an eminent astronomer, besides being the head of the government in his own city, utterly repudiated the Chaldean method of foretelling the future." (2.27)

The astrology of the day was not limited to the genethliacal kind. The astrologers predicted weather, published lists of lucky and unlucky days, foretold earthquakes, the appearances of comets, and made forecasts of events in the lives of kings and common men (2.28, 2.29). Their prognostications were not always successful:

"I recall a multitude of prophecies which the Chaldeans made to Pompey, to Crassus and even to Caesar himself (now lately deceased), to the effect that no one of them would die except in old age, at home and in great glory. Hence it would seem very strange to me should anyone, especially at this time, believe in men whose predictions he sees disproved every day by actual results." (2.30)

Pompey and Crassus were both murdered at times each had suffered military defeats. Interestingly there are other similarities between the astrology of the past and that of today. The historian Sir Gaston Maspero tells us the profession of priest-astrologer was a lucrative one, so that rivals to these Chaldeans having questionable training took advantage of the market. In Maspero's words:

"These quacks went about the country drawing up horoscopes and arranging schemes of birthday prognostications of which the majority were without any authentic warranty. The law sometimes took note of the fact that they were competing with the official experts, and interfered with their business, but if they happened to be expelled from one city, they found some neighboring one ready to receive them." (2.31)

We do not know a great deal about early Greek astronomy and astrology. There are a few references such as those quoted above, but it is clear that in Greece astrology found a receptive and innovative people and here it grew remarkably. The reason is probably a religious one. The Greeks speculated wildly about the world; occasionally they were even correct. At first the heavenly bodies were deified, and by extension the earth as well. They further reasoned that if the earth if divine, the other elements, water, air and fire must be also. The sun was worshipped as the giver of light and life. Gradually the view regarding the divinity of the seven planets was repalced by a modified picture where the planets were under the guidance of divine spirits. A vast, eternal harmony could be seen in their regular movements. There was general belief in harmony could be seen in their regular movemtns. There was general belief in prophecy and predestination. A religion that added interpretation to these signs in the sky would be well received. So astrology blossomed magically and widely in hellenistic Greece and surrounding countries.

There is sketchy evidence, mostly from secondary sources, suggesting that astrology was brought to Greece by a Chaldaean "priest of Bel" called Berossos, who set up a "school" on the island of Cos around 280 B.C. (2.32, 2.33). This seems to have become a center for astrology and medicine, the latter since Hippocrates had lived on Cos. Although there was certainly some earlier knowledge of Chaldaean practices as we have noted, astrology seems to have spread widely through Greece shortly following this time.

It is difficult to reconstruct this early history in part because so few actual horoscopes have survived the centuries (2.34). There are about 20 Babylonian cuneiform horoscopes (though about 2000 astronomical tablets) known to exist that range in dates from 410 B.C. to 69 B.C., and about 10 known Egyptian horoscopes, the earliest from 38 B.C. and the latest 93 A.D. About 200 Greek horoscopes are known (but only about 20 astronomical documents) (2.35). The dates for these range from 72 B.C. to about 600 A.D., with all but five occuring after 0 A.D. Most fall in the first two centuries A.D., owing to their preservation in several literary sources. They usually consist simply of an enumeration of the positions of the celestial bodies, and in only very few cases is an explicit forecast included, their use in an astrological connection being only conveyed through astrological literary sources. These horoscopes, in the rare instances that diagrams are given, don't look much like the charts of today (2.36). They were generally drawn in a square pattern, only few give a circular diagram (see Figure 2.3), and symbols do not appear to have been used for the planets or zodiacal signs until the middle ages.

Perhaps the most important astrological text of this time is the Tetrabiblos of Claudius Ptolemy. This collection of astrological knowledge was compiled after Ptolemy's astronomical work, The Almagest which had been written about 140 A.D. The oldest known version of the Tetrabiblos is a copy from the thirteenth century. It is interesting that Ptolemy was careful to distinguish between astronomy and astrology in his introduction to the Tetrabiblos. Mentioning two means of prediction he wrote

"One, which is first [astronomy] both in order and in effectiveness, is that whereby we apprehend the aspects of the movements of sun, moon, and stars in relation to each other and to the earth . . . We shall now give an account of the second and less self-sufficient method in a properly philosophical way, so that one whose aim is the truth might never compare its perceptions with the sureness of the first, unvarying science . . ." (2.37)

In some circles the Almagest has been considered Ptolemy's most important contribution. This is a work that encompasses the entire range of ancient astronomy, preserving observations of his predecessors such as Hipparchus, adding observations of his own, and refining the epicyclic theory of planetary motions. Surprisingly enough it has been suggested by several investigators of

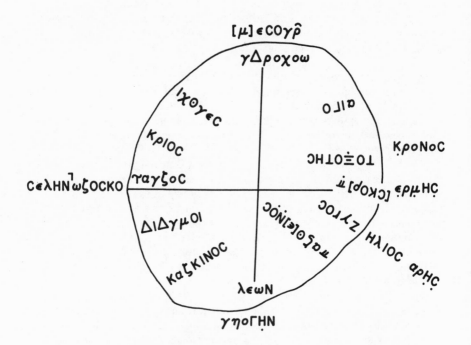

Figure 2.3

A Greek horoscope from approximately 15-20 A.D. (Neugebauer, O., and Van Hoesen, H.B., *Greek Horoscopes*, The Am. Phil. Soc., Philadelphia, 1959, p. 18.)

Ptolemy's work that a number of observations he claims to have made personally are fraudulent. Most recently R. R. Newton has presented strong evidence that Ptolemy fudged his data, that is, he calculated numbers he then claimed to have obtained by measurement, in order to make his work agree with the known work of earlier astronomers (2.38). He evidently did produce some genuine observations, but he chose not to use them.

"If Ptolemy had been content to use valid observations instead of using fudged data to support wrong hypotheses, he would have been the first astronomer to find an accurate value for the precession of the equinoxes. Distressingly, he was not. His fraudulent 'data' caused many good astronomers, including Copernicus, to waste valuable effort on the 'trepidation of the equinoxes', and consequences of his fraud can be found in astronomical research even in the present century. Indeed, the science of astronomy would be further ahead if Ptolemy had never written the Almagest." (2.39)

It is tempting to make a similar comment about the Tetrabiblos, but of course Ptolemy's possible dishonesty does not of itself mean that the astrological knowledge he has conveyed to us is untrue. However, it is difficult not to be just a bit more skeptical of Ptolemy's writings.

The middle ages brought with it a significant decline in empirical science in favor of a more philosophical approach to the advancement of knowledge. Interest in astrology, however, remained high. The fact that astronomy progressed little during this period tells us something about the astrologers, namely, that they were not "also astronomers" and they did not care about the reliability of their basic data. They had no interest themselves in the theoretical foundations of their craft or in its improvement.

"This attitude is reflected in the fact that astrologers for centuries used arithmetical methods, e.g. for planetary positions and for determining the length of daylight, which were long superseded by more accurate procedures . . . Hence one may well say that at no stage in the development of astronomy did astrology have any direct influence, beneficial or otherwise, on astronomy beyond the fact that it provided a secure market for treatises and tables and thus contributed to the survival of works which otherwise would hardly have reached us." (2.40)

Among the most presumptuous of the historical claims made by astrologers are those relating to the Scientific Revolution and its leading luminaries. In particular the astrologers claim for their own no less than Nicholas Copernicus, Johannes Kepler, Tycho Brahe, Galileo Galilei, and Isaac Newton! Unfortunately, the true attitude of these individuals toward astrology is a matter of considerable historical debate. It is Kepler who best epitomizes the confusion that exists over the matter. On one hand, in "De Stella Nova", he tells us

" . . . the excitement of sublunary natures by the conjunctions and aspects of the planets has instructed and compelled my unwilling belief . . . "

23

On the other, however, in "Tertius Interviens", he speaks of the "follies and blasphemies of astrologers", and astrology itself as "the foolish stepdaughter of astronomy . . . a dreadful superstition . . . a sortilegous monkeyplay."

In 1610 Galileo Galilei published the historic "Starry Messenger", a sensational summary of his observations at the telescope that forever changed astronomy. Astrologers fondly recall that barely one year before, Galileo cast a horoscope for his patron the Grand Duke of Tuscany. In their fondness, however, the astrologers are unable to recall that Galileo, in that same horoscope, promised the Grand Duke a long and fruitful life, to which the Duke responded by dying a few weeks later.

In a similar vein, a great many astrological texts point out with delight that the first Astronomer Royal of England, John Flamsteed, cast a horoscope for the "birthdate" of the Royal Observatory at Greenwich on August 10, 1675. Indeed, to this very day, the horoscope can be seen on display at the Greenwich Observatory. What is conveniently omitted by the astrologers, however, is the last line of Flamsteed's oft-referred to horoscope which reads "risum teneatis, amici?" (Can you help laughing, my friends?)

In an equally laughable mode astrologer Zolar, in his text "The History of Astrology", magically tranforms the Astronomers Royal of England into the "Astrologers Royal" before our very eyes (2.41). Author Ronald Davidson goes so far as to claim Albert Einstein as an astrologer on the historical basis that he "was most favourably impressed by an astrological treatise by a nephew of the late Dr. Adler . . . " (2.42). Astrologer Robert Jansky (2.43) has Hippocrates as a "student of Ptolemy", despite the fact that the good doctor predates Ptolemy by some six centuries. Such are the bases upon which the astrologers lay claim to various historical figures.

Ironically, in resorting to such tactics the astrologers have created a most interesting historical paradox for themselves. If indeed the great thinkers of the Scientific Revolution embraced astrology to the extent that astrologers would have us believe, then why did the 17th and 18th centuries witness a *decline* in astrology at the precise instant in history when all of the sciences from astronomy to zoology were in the ascendant? The answer is very simple. The great scientific thinkers of the day gave astrology a chance to demonstrate its effectiveness as a possible vehicle with which to probe the secrets of the physical universe. When the astrological vehicle was found to suffer from continuous breakdowns, the framers of the Scientific Revolution rejected it out of hand. With the possible exception of Kepler, no scientist after the turn of the 17th century can be thought of as accepting astrology in any way during their later years. To claim these individuals for astrology, therefore, is comparable to the Catholic Church claiming for its ranks Martin Luther and Henry VIII of England on the basis of their early membership in that body. The late Senator Joseph McCarthy made a career of branding as "communist" any individual who had *any* flirtations of any sort at any time of their lives with that ideology. Included in his list of "communists" for example, were Adlai Stevenson and Edward R. Murrow. One cannot help but note the use of similar tactics on the part of the astrologers in attempting to advance their historical claims.

Of course, it has not been our intent here to even attempt a complete outline of the historical development of astrology as compared with astronomy. It would nevertheless be of further interest to extend this summary briefly to mention astrological endeavors in the Far East. Astrology, if anything, has had a far greater following and impact in India and China than in the west. Its history is certainly as long, having begun in China more than 4000 years ago.

Although there are some similarities there are many significant differences between eastern and western astrology (2.44). For one thing the pole star was important to the Chinese because, like the emperor around whom the kingdom revolved, it too was a focal point. Thus much attention was directed toward the circumpolar stars which could be seen year round and through the whole night. This portion of the heavens was divided into eastern, western, northern, and southern quadrants or "palaces". Around the equator there were 28 "constellations" called "hsui" or mansions. The 28 include the Swallow, Tapir, Worm, Monkey, and Hare for example, and bear no relation to western constellations or zodiac. Some of the same animals appear in the oriental "circle of animals" which lends its names to the twelve months, and a cycle of twelve years. The menagerie begins with the Rat, and continues with the Buffalo, Tiger, Cat, Dragon, Snake, Horse, Sheep, Monkey, Cock, Dog, and Pig. Chinese horoscopes are further different in ignoring the ascendant, a point considered very important in the western version. They did note the moon's conjunctions with stars and planets and its altitude, and likewise the sun and planets and their colors. And as the Greeks had a quartet of elements, the Chinese had five, adding wood to fire, earth, water, and metal, and associated them with the bright planets.

Hindu astrologers also watched the passage of the moon and divided the zodiac equally into 28 nakshatras or mansions with portents for all. There was greater communication with Greek astrology here and so there are more similarities. Several tropical and sidereal zodiacal systems have been in use in India. Also here there is a much closer religious tie than in the west, partly accounting for the wider acceptance of astrology.

The national and cultural difference in astrological systems are typical. Even within one cultural group there may be competing procedures espoused by one group or another giving quite divergent predictions. The physical sciences stand in sharp contrast to this. For whereas astrology in China may scarcely resemble astrology in the United States, their sciences are quite equivalent. The physics employed in China is precisely the same physics that is applied in the United States. The identical physics is used in India, and the Soviet Union, and Saudi Arabia. It is also the same physics by which the distant stars shine.

2.1 Jones, M. E., p. 37, "Astrology: How and Why It Works", Shambhala Publications, Inc., Boulder, Colorado (1969).

2.2 Lewi, G., p. 1, "Astrology for the Millions", fourth revised edition, Llewellyn Publications, St. Paul, Minnesota (1969).

2.3 Pannekoek, A., p. 19, "A History of Astronomy", George Allen S. Unwin LTD, London (1961).

2.4 Baver, T., Zeit fur Assyr. und Asiat. Arch., 43, 308 (1936).

2.5 van der Waerden, B. L., "Science Awakening II: The Birth of Astronomy", Oxford Univ. Press, New York, New York, p. 49.

2.6 Ibid., p. 49.

2.7 Reiner, E. and Pingree, D., p. 29, "Babylonian Planetary Omens Part I Enuma Anu Enlil", Undena Publications, Malibu, California (1975).

2.8 Ibid., Reference 2.5, p. 59.

2.9 Biggs, R. D., p. 28 "SA.ZI.GA Ancient Mesopotamian Potency Incantations", J. J. Augustin Publisher, Locust Valley, New York (1967).

2.10 Leichty, E., "The Omen Series Summa Ifbu", J. J. Augustine Publisher, Locust Valley, New York (1975).

2.11 Tyl, N., p. 4, "The Principles and Practice of Astrology", Llewellyn Publications, St. Paul, Minnesota (1974).

2.12 Ibid., Reference 2.10 p. 32.

2.13 Ibid., Reference 2.10, p. 33.

2.14 Ibid., Reference 2.10, p. 42.

2.15 Ibid., Reference 2.10, p. 75.

2.16 "Cuneiform Texts from Babylonian Tablets in the British Museum", Part V, Oxford Univ. Press, London (1898), CT5, plate 4. Translation; A. Guinan, Ann Asinan, private communication.

2.17 Ibid., Reference 2.16, CT40, plate 48.

2.18 Ibid., Reference 2.16, CT38, plate 36.

2.19 Hays, H. R., p. 68, "In the Beginnings", A. P. Putnam's Sons, New York, New York (1963).

2.20 Durant, W., p. 244, "The Story of Civilization: Our Oriental Heritage", Simon and Schuster, New York, New York (1954).

2.21 Ibid., Reference 2.5, p.

2.22 Sachs, A., J. Cuneif. Studies, 6, 54 (1952).

2.23 Ibid., p. 57.

2.24 Ibid., p. 65.

2.25 Diodorus, Book II, 29.

2.26 Strabo, XVI.I, "Geography", Loeb Classical Library, Harvard University Press, Cambridge, Massachusetts (1940).

2.27 Cicero, Bk. II, 99, "De Divinating".
2.28 Rawlinson, G., p. 579, "The Five Great Monarchies of the Ancient World", Vol. II, John Murray, London (1879).
2.29 Ibid., Reference 2.27, 31.
2.30 Ibid., Reference 2.29, 99.
2.31 Ibid., Reference 2.30, p. 780.
2.32 Ibid., Reference 2.5, p. 113.
2.33 Neugebauer, O., p. 607, "A History of Ancient Mathematical Astronomy", Apringer-Verlag, New York, New York (1975).
2.34 Neugebauer, O. and van Hoesen, H. B., p. 161, "Greek Horoscopes", The American Philosophical Society, Philadelphia, Pennsylvania (1959).
2.35 Ibid., p. 162.
2.36 Ibid., p. 163.
2.39 Ptolemy, C., p. 1, "Tetrabiblos", Loeb Classical Edition, Harvard University Press, Cambridge, Massachusetts (1940).
2.38 Newton, R., Qtryly. J. Roy. Astron. Soc. *14,* 386 (1973).
2.39 Newton, R., Qtrly. J. Roy. Astron. Soc. *15,* 120 (1974).
2.40 Ibid., Ref. 2.33, p. 943.
2.41 Zolar, p. 252, "The History of Astrology", Arco Publishing Company, Inc., New York, New York (1972).
2.42 Davison, R., p. 13, "Astrology", Arco Publishing Company, Inc., New York, New York (1963).
2.43 Jansky, R., p. 6, "Astrology, Neutrition and Health", Para Research, Rockport, Massachusetts (1977).
2.44 Gleadow, R., p. 86, "The Origin of the Zodiac", Atheneum Press, New York, New York (1968).

Chapter 3
The Shadow of Doubt

No one likes to be wrong. Yet we are all wrong for one reason or another from time to time, when we were quite confident we were right. We may have argued with some vehemence, even wagered, certain the other guy was wrong. But *we* were wrong. Perhaps we had misremembered something, or maybe we relied on some authority, some source of information that later turned out to be quite erroneous. It may have been very embarassing, even costly. We were so *sure*. But it can happen; it's all part of being human and a little less than perfect. After all, it is very difficult at times to know what the facts are and to make the right decision. It is a very complicated world out there — lots of shades of grey, vested interests, incomplete information, calculated risks. You have to be very, very careful. Right? Right.

Well, what *can* you believe? It is a key question; one that is central to everything we are going to discuss in subsequent chapters. How do you decide for yourself whether or not astrology works and if the planets affect the course of human events? How do you know that astronomy, or anything else, is correct? These are deceptively simple questions, and the answers are not clear cut. Through all of history no one has uncovered any magic formula that can be applied to yield a result incontrovertably true. There in fact seems no infallible way of deciding what is "truth". Though this may be an unhappy situation, it is not a hopeless one. For we believe there is very definitely a *best* way to go about finding what is most likely to be right. It is a procedure that has been forged over the centuries as part of the progress of mankind in philosophy and science. Though not easy to fully define, in principle it is simple. It is the approach to knowledge gathering that has been called the *scientific method* or the *method of hypothesis testing*.

Briefly the scientific method is a sort of cyclic trial-and-error approach to the

28

investigation of nature. Here, accurate experimental results and carefully observed phenomena are the only "truths". The observational results are summarized to reveal behavior patterns or laws which in turn are collected into a very general description of the world (see Figure 3.1). While this might have been good enough at one time, the modern scientist is neither content nor permitted to stop here. The theory must be used to make experimentally verifiable predictions which can then be subjected to appropriate tests. If the prediction and the experimental check agree, the theory is shown to possibly not be incorrect. So the scientist must continue to make more testable predictions based on the theory. Sooner or later the prediction and observation may not agree. At this point the law or theory being tested may be discarded or at least amended in order to account for the predictive failure. More predictions are then made using the new theory. And so on.

This summary of the scientific method is of course not the whole story. Any individual scientist does not necessarily follow all of the steps or the order of the above pattern. He might just formulate a theory based on the experimental results of others. Or he might only be performing one of the many tests of the theory that will be carried out. Certainly many experimental tests are exceedingly delicate, difficult to perform, and technically complex, so talented specialists may be needed. Furthermore scientists are not the dispassionate, logical investigators of fiction or cinema, but creatures subject to the same emotions as everyone else. But in any case, it is a procedure that clearly works. About this there is no doubt. No bit of established knowledge about the world around us or of modern technology has come about in any other way but through the scientific method.

Our description raises other questions. For example we might ask why, if the prediction and experiment agree, does that not prove the theory to be correct? Why do you have to go on testing indefinitely? Can't you just identify some point where it has been "proven"? Sadly, no. As it happens you can never prove anything beyond question no matter how good the agreement between theory and experiment. This is an important limitation that comes from the fundamental logic used by science. It is such a vital part of what we are doing that we shall discuss it in some detail.

The scientific method in part may be viewed as a kind of logical argument. We are trying to defend or disprove some description of nature. Formally an argument is just a group of statements that have some sort of relationship to one another. Someone might argue with you for example that your 23 year old pet cat won't be around much longer since cats don't live forever. Not much of an argument in the usual sense, since you are likely to agree with your friend. But note that there are several parts to this line of reasoning that can be identified here if we wish:

> statement 1: "all mammals are mortal"
> statement 2: "all cats are mammals"
> statement 3: "all cats are mortal"

Statements 1 and 2 are the premises of the argument, statement 3 is the conclusion. This is a common form of argument in which the conclusion can be

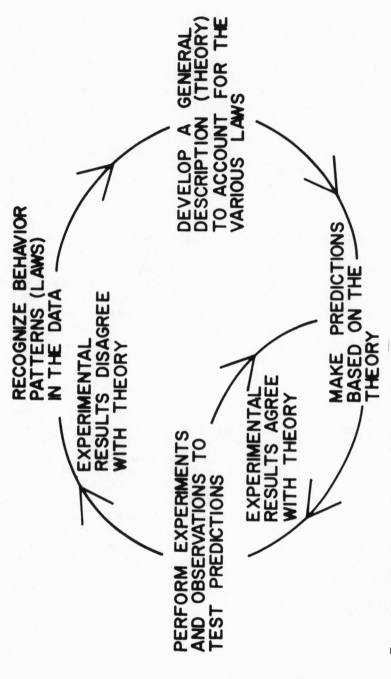

Figure 3.1 A schematic diagram of the Scientific Method.

deduced from the premises.

Logic may be defined as that part of philosophy which deals with how premises and conclusions are related to one another (3.1). If the argument is logically correct it is said to be a *valid* argument. Note that the logical correctness has nothing at all to do with whether or not the premises or conclusions are true or false, but only with how they are related to each other. And this relationship for a valid argument is such that if the premises are true the conclusion will always be true. You may on the other hand have false premises and a false conclusion and still have a valid argument:

> All moons are green cheese
> All planets are moons
> All planets are green cheese

The *form* of this argument is identical to the first one, so it is logically valid; the form alone determines this. An invalid argument differs from a valid one in that even though the premises are true, it does not follow that the conclusion is true. It could just as well be false. You cannot tell from the argument.

It is more likely that we shall encounter arguments that contain conditional statements. A conditional statement is one like "If Smithers was born on June 13, then he is a Gemini". This statement has two parts, the "antecedent" clause, the part following the "if", and the "consequent" clause after "then". In a condensed notation it has the form "If A, then C".

For much of history it was thought that the earth was in the center of the universe (Astrology still places the earth at the center of things). Everything appeared to circle the earth so it was a reasonable first attempt at understanding how the celestial realm was organized. There is a basically simple experiment that can be performed to test this notion. That is, if the earth were being circled rather than itself circling the sun, you should not see any apparent shift (parallax) of a star that would arise if the earth were on opposite sides of the sun every six months. If you were to go out and measure with a sufficiently precise instrument the apparent direction of a nearby star, you will find however that it in fact shifts very slightly back and forth with the swing of the earth around the sun in its orbit. The rational choice is then to reject the theory which says that the earth is stationary. We can summarize these statements in the form of a conditional argument.

(Arg 1)
> If the earth is stationary, then the stars will show no apparent shift.
> But observations show the stars do exhibit an annual shift.
> Therefore it is not true that the earth is stationary.

This argument has the form

> If A, then C
> C is not true
> Therefore A is not true

This is a valid form of the deductive conditional argument and is called "denying the consequent". If the premises are true, the conclusion will also be true. In other words when you venture forth to compare theory and prediction with observations, and the observations or experiment show the prediction to be incorrect, you are compelled by the strength of logic to reject the theory, and begin again.

Often our attempts to empirically match a theoretical prediction will be successful. We might revise our theory above to incorporate the knowledge gained from the failure of our initial attempt to now claim that the earth goes around the sun.

> If the earth circles the sun, then the nearer stars will show an
> apparent annual shift.
> (Arg 2) The observations show the stars do exhibit an annual shift.
> Therefore the earth does circle the sun.

Now everything fits very nicely for it seems that we have proved our new theory correct. And this is precisely what we do in experimental science: confirm theoretical predictions by observational test. This kind of deductive conditional argument is called "affirming the consequent". There is just one little problem with the argument. It is invalid.

We must keep in mind that the validity of an argument depends only on its form. The fallacy of affirming the consequent can be put into our symbolic notation as:

> If A, then C
> C is true
> Therefore A is true

An argument that is obviously nonsense but that exactly conforms to the pattern of the argument can illustrate its invalid nature.

> If the University of Virginia is in Georgia, then U. Va. is in the south.
> (Arg 3) The University of Virginia is in the south.
> Therefore the University of Virginia is in Georgia.

Logically there is no difference between these two arguments (2 and 3). Each is equally invalid. It is of no matter whether the statements in the argument are true or not, it does not follow that the conclusion is true. Thus in every case that an experimental result is consistent with what was expected from theory or hypothesis, nothing is *proven,* no matter how good the agreement between theory and fact. The observation may be correct or the experimental result accurate, but in no way does the argument require that it follow from the particular hypothesis in question. The experimental result might just as well follow from some other completely different description of nature. In the example above (Arg. 2), a more imaginative person might propose that the stars are of themselves shifting back and forth over the course of a year. The single observation of phenomenon does not permit you to distinguish between the two hypothesis; future information might, however, make it clear which is the better choice.

It is for this reason that science can never rest content that it has found the answer (even though the individual may have faith that he is right). Thus in

32

science, having produced some description or explanation for something in nature through some creative flash of insight, the hypothesis must be submitted to every possible test (within practical limits) and compared with all available relevant data. It is especially useful to devise novel and unusual tests for what they might reveal. Occasionally it is a new set of circumstances that shows the limits and incorrectness of the hypothesis (e.g. the failure of Newton's laws of motion at velocities approaching the light). There may even be a certain satisfaction in the definiteness of the progress made when you are able to show a given hypothesis does not fit the observations (recall that denying the consequent is valid). It is often not fully appreciated, especially by those at the fringes of science, that one of the roles of the scientist is that of adversary, trying to falsify a particular hypothesis. Philip James Bailey has written,

"Who never doubted never half believed
Where doubt there truth is — 'tis her shadow" (3.2)

For only if ideas survive the most penetrating criticism possible, can you begin to raise your hopes that some additional truth has been found. Then as more favorable evidence accumulates, hence the more chances that passed when the idea could have turned out to be wrong, the higher the likelihood that it is right. However, though the probability increases toward a practical certainty, it can never reach 100 percent, and there is forever and always some chance of being wrong.

So science says try it, compare it to the real world; if it does not fit, *throw it away.* If it does seem to fit, it may be right, but you must keep on checking and checking, because science can never be absolutely sure. And if science remains uncertain, can you be so egocentric and arrogant as to believe you know the truth?

The results obtained from observation or experiment make up the evidence upon which we base our judgement of whether or not the hypothesis under test is confirmed. Our conclusion must depend not on our personal biases regarding the hypothesis, but on the completeness and quality of the evidence. There is good evidence and bad evidence. First the bad evidence. Some of the poorest evidence is the kind we gather with our senses and the impressions we have of the way things are. For example, if you place two pans of water into a freezer, one hot and one cool, which of these does your intuition tell you would freeze first? It is in fact not the cool one but the hot water that freezes most rapidly (3.3). And is it not intuitively plausible that a heavy rock should fall faster than a tiny pebble? Yet it can easily be verified that a rock and pebble dropped together hit the ground at the same instant.

We may think we view the world without error. Have you ever seen an optical illusion, or misidentified a stranger who looked like someone you know? Many people are color blind, or near-sighted. Count the number of people in a crowd wearing glasses sometime. And we all have heard stories or seen demonstrations of how inaccurate witnesses at an accident can be. People have been known to have delusions and hallucinations. Have you had your hearing checked lately, or ever heard any bells or whistles that weren't there? Fortunately for our sanity we tend not to remember unpleasant things. Other times we may select those things which tend to support our opinions for later recall as "incontrovertible" facts. Sometimes we simply forget. We are, in short, inaccurate and unreliable observers of the world.

Lord Kelvin once wrote that:

" ... when you can measure what you are speaking about and express it in numbers you know something about it; but when you cannot measure it, when you cannot express it in numbers, your knowledge is of a meager and unsatisfactory kind; it may be the beginning of knowledge, but you have scarcely, in your thoughts, advanced to the stage of science ... " (3.4)

Stepping outdoors on a summer afternoon, you may have the impression that it was warmer than the day before. But if the thermometer reads 33°C and yesterday it read 35°C, you in fact know that it is not hotter today. Quantitative data, the good evidence, is the foundation of scientific advancement. Technology allows us to weigh it, measure its length, or read it from an LED display with ever increasing accuracy. We must also know something about the properties of these instruments to judge the quality of the data. Often quantitative information may take the simple form of a count. For example to find out on which day of the week the traffic on Main Street is heaviest, stand there and count the number of vehicles passing each day. It is clear that you must do a good job of counting. In other cases you may not be able to or need to count everything, a sample of data may suffice. If the traffic flow were constant during a given day you could just count for one hour and multiply by 24 to have a good estimate of the total number of vehicles. The kind of sample depends upon the problem. Pollsters do quite well with a "representative" sample of about 1500 people to statistically reflect the opinion of the nation as a whole. Whatever it is then, if you have numbers to compare, data that accurately mirror what the real world is like not what you think it is like, you are on the right track to make an informed decision.

As in the laboratory you must know your instruments, statistically you must know the properties of your data sample. It is important to avoid selecting just the data that fits. If an astrologer tells you that critical events occurred in your life at ages 14, 21, and 26, and you recall that at 14 you moved to Colorado, and at 21 you graduated from college, and at 26 you were married, you might be impressed with the evidence of success. But how many other things happened to you of varying degrees of significance each week of your life? If age 23 were mentioned could you not have recollected something that happened that year? You must count up the misses as well as the hits. In this sort of comparison your "data" must be adequately complete and not limited in such a way as to bias the outcome.

There are some less tangible criteria that in a very general way have bearing on the conformation and acceptability of our descriptions of the world. Any alleged explanation for any phenomenon or series of events may be reasonably expected to have some relevance to the phenomenon if it is to be of any value. Consider the following. While watching a total eclipse of the moon on a spring evening in 1968, one of us noticed that several people came out of an apartment complex and began to bang on some pots and pans. Interestingly a short time after this display, the moon began to reappear and was shortly back to its usual glory. We could propose the following explanation for the sequence of events:

At a total lunar eclipse, the moon reappeared when several people made a lot of noise by thumping on pots and pans, thus the noise saved the moon from a horrible fate.

Having left the dark ages far behind, and knowing the moon's emergence from the earth's shadow was inevitable (barring unimaginable disaster), we realize the noise hypothesis has nothing to do with the lunar eclipse and so is quite useless, though perhaps amusing. In spite of the evidence showing the eclipse ended apparently after the noise making commenced, we know on other grounds that the hypothesis is absurd. It lacks what has been called *explanatory relevance* (3.5). It's a bit like claiming credit for household safety from having daily pronounced "Bless this house and keep it safe from tigers" when you live in New Jersey. Certainly it is not always possible to make such a pre-judgement from established knowledge, but when it is we are well advised to discard the irrelevant explanation without further ado.

We might in this regard be faced with two hypotheses suggesting the origin of our own individual characteristics, one from astrology and one from science. Astrology says that our personality attributes are determined from the pattern of planets in the sky at our moment of birth. Sciences says that our personality attributes are determined from the combined genes of our parents and our early environment. Which of these hypotheses has explanatory relevance? Which one should we choose to retain?

In a similar vein, the credibility of any hypothesis which conflicts with well-established current knowledge will always be quite low, at least initially. Firmly established knowledge has the weight of extensive and successful experimental testing behind it, whether correct or not, and so is not lightly tossed aside. Thus any radical hypothesis is likely to require exceptionally strong evidence to attain any degree of acceptance. It is not that science wishes to preserve its favorite ideas against adverse evidence, quite the contrary, as the history of science and its revolutions show. However it is legitimate to raise objections on theoretical grounds within reason. This creates a system of checks-and-balances within science that assists in screening the useless from the valuable hypotheses. There is scarcely the time to devote to checking every single suggestion of every imaginative person. So if applied with judicious constraint, one hopes that nothing useful is lost, and resulting progress may be more rapid. The scientist, and every rational being, must be the eternal skeptic, for he knows only too well how easy it is to be wrong.

It should also be clear that for any hypothesis to be worthy of serious consideration, be it some precept of astrology or the theory of stellar evolution, it must be open to objective empirical test (3.6). For if there is no imaginable observation or experiment that can be done as a check, then it will be impossible for the "theory" to either agree or conflict with anything. In some cases practical difficulties may arise with such empirical testing. Who, for example, would be willing (or able) to watch stars evolve over several billions of years to check on a theory? However, we may gather admissible evidence for our trial through computer simulation of the physical situation. Such modeling of complex physical and biological systems has become an exceedingly valuable technique in the physical sciences, engineering, economics, and so on. In any case if an idea cannot even be at least tested "in principle" it can have no bearing on any observable phenomenon. That is, suppose someone claims that the planet Venus affects the growth of plants in a certain way through "vibrations" or emanations produced at particular times, especially in the spring. The effect of these emanations is that the plants produce flowers. Furthermore these emanations have the property that they cannot be detected by any conceivable device (which is why no one has discovered them before) nor can any biological effects be found in the plants, except that the flowers happen to bloom in exact synchrony with the moment the emanations arrive from Venus. Buds may form, but they could never bloom except in response to these emanations. Now we have built into this hypothesis attributes that exclude its testability. There is positively no way to tell if these rays are there or not. Then you cannot tell if the hypothesis is right or wrong. The emanations do not exist. The hypothesis, lacking testability, is worthless, and we have excellent grounds for rejecting it. And we should, since to accept it would be self-delusion.

To further illustrate our point regarding the testability of hypotheses, let us take an astrological example. Modern astrologers have assumed a statistical stance in their approach to indicators of daily events in the horoscope. The fulfillment of any given prediction is not required, since it has been claimed since Ptolemy's time, that "the stars incline, but do not compel". We will confine ourselves here to one comment on this principle apropos of the current discussion. Taken to its ultimate, this principle effectively removes this part of astrology from the empirical realm. Given that you may be expected to lose your temper whenever Mars is square Pluto, if it *never* happens, and the astrologer then invokes incline-not-compel, astrology is protected and cannot be disproved. By this there cannot be a disagreement between prediction and experiment. Then by the requirement of testability, astrology would be without force in the world.

Implicit in much of this discussion and throughout science, is the notion of causality, the idea that given the same set of conditions, they will be followed every time by the same event. In other words, there is an identifiable sequence of interelated events, frequently simultaneous, linking some definable first event with a consequent phenomenon of observable resultant events. Same cause, same effect. There is a certain sense of security in such determinism, but a number of philosophical problems as well. It is widely known that quantum theory has encouraged a somewhat different viewpoint wherein there is not a strict cause-effect relationship between events in the real world, at least on the microscopic or atomic level. Astrologers, having recently recognized this, believe they have found support for the view that no physical connection need exist between the planets and people for astrology to work. However, there is little solace for astrology here. It is true on the quantum level that individual events are not predictable, but one can establish probabilities for given events and see the collective result fulfilled for many events. Astrology, after 3,000 years, doesn't seem to have yet reached this stage as we shall show later. The quantum level effects, at any rate, do not extend upwards to the scale of the solar system. Furthermore, there are philosophers and physicists that support a strict causal connection at all levels of nature for all events, saying that we only do not yet have all the information needed to express this relationship (3.7). In any case criticism of causality is not new. It goes back to the 18th century and David Hume who argued against the necessity of strict causality. Subsequent critics also included Bertrand Russell who saw sequences of events as expressing some functional dependence. Nevertheless, it still seems that a large part of the search for explanations of things in the real world is usefully described as the identification of "plausible physical relationships" between events. Offered the choice between the hypotheses that you feel depressed today because Neptune is transiting your sixth house and you feel depressed today because of the low barometric pressure, we are likely to not consider the former very seriously owing to the lack of physical connection. This does not "prove" that the first hypothesis is wrong. Indeed there is no established discipline that rejects causality as completely as astrology (at least some astrologers). This has been a major factor in preventing its acceptance, and we shall feel inclined (but not compelled) later to examine astrology for any plausible physical relationships.

In summary we would say again that science is the best way to find truth known to mankind. It is fair and objective, even if scientists are not; we certainly don't claim to be totally free of bias. This is partly why the scientific search for truth is so important. It transcends bias and personal opinion, rejects merely psychologically satisfying explanations, and produces evidence for anyone to see and judge. It is not limited to "hard" science. Abraham Maslow has advocated the broadest application of science, writing that science

" . . . need not abdicate from the problems of love, creativeness, value, beauty, imagination, ethics, and joy, leaving these altogether to 'non-scientists', to poets, prophets, priests, dramatists, artists, or diplomats. All of these people may have wonderful insights, ask the questions that need to be asked, put forth challenging hypotheses, and may even be correct and true much of the time. But however sure *they* may be, they can never make mankind sure. They can convince only those who already agree with them, and a few more. Science is the only way we have of shoving truth down the reluctant throat. Only science can overcome characterological differences in seeing and believing. Only science can progress." (3.8)

The evidence – objective descriptions of nature – is the only basis for truth. It is not what we think the world is like, it is what the world reveals to us whatever our preconceived ideas. To believe without evidence is imagination; to believe in spite of the evidence is delusion. Only empirical evidence counts. Whatever criticism and objections may be raised, the "proof" of the idea must be that it works. We must demand of any hypothesis that it passes every empirical test we can pose for it. We demand no more or less from astrology.

3.1   Salmon, W.C., p. 4, "Logic", Second ed., Prentice-Hall, Inc.; Englewood Cliffs, New Jersey (1973).

3.2   Bailey, Philip James, "Festus: A Country Town" (1839).

3.3   Kell, G.S., "The Freezing of Hot and Cold Water", Am. J. Phys. 37, 564 (1969).

3.4   Kelvin, William Thompson (Lord), "Popular Lectures and Addresses", (1891-94).

3.5   Hempel, C.A., p. 48, "Philosophy of Natural Science", Prentice-Hall, Inc., Englewood Cliffs, New Jersey (1966).

3.6   Ibid., p. 30.

3.7   Schlegel, R., "Historic Views of Causality", Causality and Physical Theories, AIP Conference Proceedings No. 16, W.B. Rolnick, Ed., American Institute of Physics, New York (1974).

3.8   Maslow, A.H., p. viii, "Toward a Psychology of Being", Second ed., D. Van Nostrand Co., New York (1968).

Chapter 4

The "Librarians" Universe

In one of his more ebullient and presumptive moments on national television, Sydney Omarr once referred to astronomers as "the librarians of the astrologers" who went out and got their own science.* We have already examined the myopic view of astronomers as librarians of astrologers in Chapter 2, but Mr. Omarr is quite correct, however, when he tells us that astronomers have built their own science. Indeed, the development of the science of astronomy over the past three centuries has been one of humanity's most profound and exciting journeys. It also provides us with a most fascinating contrast between the techniques of knowledge gathering employed by the astronomer and astrologer.

The contrast is best developed using the works of Claudius Ptolemy as our starting point. As we have seen, Ptolemy wrote compendia of the state of knowledge in both astronomy (*The Almagest*) and astrology (*The Tetrabiblos*) around 140 A.D. For over a dozen centuries afterward, both astronomy and astrology languished in the doldrums of medieval semi-science, and as a result, on the eve of the Scientific Revolution, there was very little in either Ptolemy's *Almagest* or *Tetrabiblos* which seriously clashed with 15th century knowledge.

The 15th century view of the physcial universe was thus, essentially Ptolemaic in nature. The earth occupied the center of the system and was in turn orbited by the sun, moon, and planets (see Figure 4.1). To account for the various pecularities of the planetary motions such as retrograde motion,** Ptolemy and his Hellenistic predecessor Hipparchus of Nicaea developed a model which employed geocentric orbits (deferents), secondary orbits (epicycles) and other

*In Search of . . . Astrology," NBC Network, January 15, 1978.

**A phenomenon in which a planet appears for a few weeks to move in a "backward" or retrograde direction relative to the background stars.

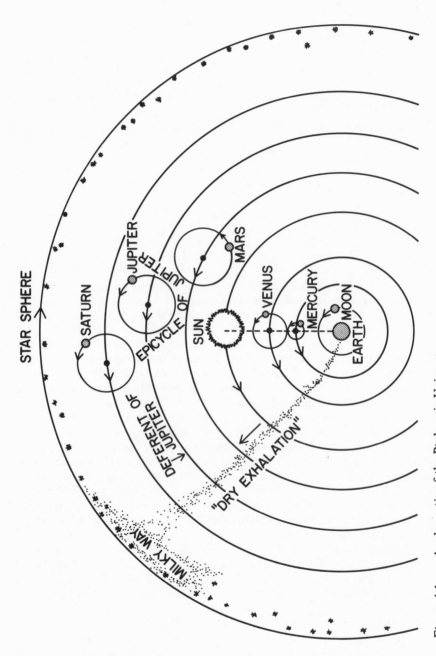

Figure 4.1   A schematic of the Ptolemaic Universe.

40

forms of geometric ingenuity. The outermost sphere in the system was the so-called "star sphere" which was occupied by all of the "fixed" stars and constellations. The Milky Way, Magellanic Clouds, and other diffuse patches of light scattered about the heavens were thought to emanate from the earth as "dry exhalations" which then rose up to take their places on the star sphere.

Here was a view of things that was satisfying to all concerned parties. For both astronomer and astrologer the system provided the means by which solar, lunar and planetary positions could be predicted for any particular time. It was a particularly attractive system for the astrologer in that the various constituents of the heavens which were deemed to be astrologically significant were also located close to the earth and would hence have no difficulties in the exertion of their astrological powers. Even the Church, that stalwart of medieval society in Europe, had few problems in accepting the Ptolemaic system. After all, what more appropriate place for the greatest of God's creatures than at the center of the universe.

The seeds for the destruction of the Ptolemaic system, however, had been in existence for centuries. Aristarchus of Samos, in the third century B.C., had suggested the idea that the sun and not the earth was the center of the planetary system. The long-dormant idea found renewed life in the mind of the Polish astronomer-mathematician Nicholas Copernicus who sought a simpler system by which celestial positions could be calculated. In 1543, Copernicus published his famous *Concerning the Revolutions of Celestial Spheres* in which he expounded on the theoretical and computational virtues of a heliocentric planetary system. The real mauling which was ultimately to be administered the Ptolemaic system, however, would not begin until several decades later.

In 1609, the Florentine Galileo Galilei heard of a new Dutch invention, the telescope, by which distant objects could be made to appear closer to an observer. In characteristic Renaissance fashion, he constructed his own instrument and then he turned it skyward. The sights he saw rocked all of medieval astronomy. He found spots on the sun, mountains, craters, and valleys on the moon, and satellite "stars" orbiting the planet Jupiter. He found that Mercury and Venus exhibited changes of shape, or phases, just like the moon, and that the Milky Way was composed of a myriad of individual stars. Interestingly, the stars retained their star-like images when viewed through the telescope, and Galileo immediately recognized this as being due to the fact that these objects were far enough away to elude the resolving ability of his telescope. Thus, came the first direct observational evidence that the stars might lie well beyond the boundaries of our planetary system.

It was Galileo's telescopic observations of the sun, moon and planets, however, that confirmed in his mind the validity of Copernicus' heliocentric model. The existence of a complete set of phase changes for the planets Mercury and Venus combined with their inability to be separated by more than 27° and 47° from the sun, respectively, could be accounted for only if both of these worlds orbited the sun, and not the earth. The discovery of small satellites orbiting a larger Jupiter coupled with the fact that a smaller moon orbits a larger earth, thus cast considerable doubt on the Ptolemaic claim that the sun, which was known to be larger than the earth, orbited the earth instead of vice versa. Responding to these and other observational considerations, Galileo wrote *A Dialogue Concerning the Two Chief World Systems* in which he went a step farther than Copernicus by claiming that the heliocentric model was not theory but scientific fact. For over fifteen years, Galileo delayed publishing *The Dialogue* and when he finally went to press in 1632, he almost immediately found himself before the Roman Inquisition. In one of the most infamous incidents of history, Galileo was forced to renounce his belief in the "foolish and absurd" idea of a heliocentric planetary system. However, such antics did little to stall the ascendancy of the Copernican system. With the help of the German mathematician Johannes Kepler, who replaced the Copernican circular orbits with ellipses, the heliocentric view of the planetary system had gained almost total scientific acceptance by the end of the 17th century. Indeed, this refined view of the solar system contributed heavily to the ultimate formulation of Newton's classical theory of motion and gravitation in 1680.

Ptolemy's seven "planets" have, of course, experienced a fair share of scientific scrutiny. Not only has the sun taken over as master of the planetary motions, but it is also recongized to be the source of virtually all of the light by which the moon and planets glow in our night sky. Moreover, fossil records on the earth indicate that the sun has been radiating energy at or near its present level for hundreds of millions of years (4.1). The mechanism by which so much energy could be radiated over such a long period of time eluded scientists until the present century. The sun is now known to create its vast outpouring of energy by fusing hydrogen atoms into helium atoms at its center where the temperatures are of the order of fifteen million degrees Kelvin. In the process some of the hydrogen matter is converted into its equivalent energy according to the famous Einsteinian equation:

Energy released = mass x speed of light squared

or in a more familiar form, $E = mc^2$. It is this same energy generation scheme that has placed in our hands the ability to annihilate thousands of years of human development in a few short minutes.

In the past three hundred years, astronomers have also discovered three additional major planets, Uranus, Neptune, and Pluto, and large numbers of lesser bodies including some 33 satellites orbiting the major planets, numerous comets and asteroids, and interplanetary gas and dust. There are even sets of rings of cosmic debris known to surround the planets Jupiter, Saturn and Uranus.

In recent years, powerful radar telescopes and data transmitting space probes have enhanced our perception of the moon and planets even further. The surfaces of Mercury, Venus, Mars, and the moon have been skillfully mapped from afar, and dozens of kilograms of lunar material have been returned to the earth for scientific examination. Interestingly, the chemical elements composing the lunar soil have been found to be identical to those which make up the terrestrial soil upon which we stand. Among the scientifically discovered wonders of the solar system, we can include a volcano on Mars called Olympus Mons which has a base equal in area to the entire state of Colorado and towers

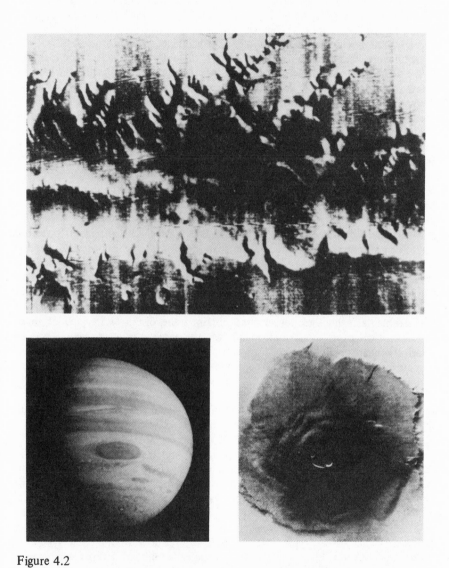

Figure 4.2

Some modern scientific views of the solar system: Top: Valles Marineris (Mars). Bottom left: The Great Red Spot (Jupiter). Bottom right: Olympius Mons (Mars).

over 20,000 meters above the surrounding terrain.* Valles Marineris, the Great "Grand Canyon" of Mars would stretch from coast to coast if superimposed on the Contenental U.S. and would plunge 6 kilometers into the earth. The earth's Grand Canyon in northern Arizona would thus, be but a tiny tributary canyon compared to Valles Marineris. There also exists an oval-shaped cyclonic storm in the atmosphere of Jupiter, called the Great Red Spot which is 40,000 km long and 16,000 km wide and has been raging for well over three centuries!

Despite the fact that by the end of the 17th century the earth's true location in the solar system had been established, astronomers found that there were still a few loose ends to clean up. One of the age-old objections to the heliocentric theory was the failure of observers to detect the so-called "parallax" effect. If the earth orbits the sun, then nearby stars should appear to shift their position relative to the more distant background stars (see Figure 4.3). When the ancient observers failed to detect such a shift, they concluded from this evidence that the earth was not moving, but rather was stationary and hence had to be located at the center of the planetary system. The triumph of the heliocentric system in the 17th century left astronomers with an orbiting earth and still no observable parallax effect. Since the amount of a star's parallactic shift decreases as the distance to the star increases, a way out of the dilemma is to assume that the stars are located at enormously large distances from the solar system. But how far? For over two centuries after the invention of the telescope, astronomers struggled to detect the elusive stellar parallax effect. Finally, in the late 1830's, three astronomers, Friedrich Bessel of Germany, Friedrich Struve of Russia and Thomas Henderson in South Africa almost simultaneously measured parallactic shifts, respectively, for the stars 61 Cygni, Vega, and Alpha Centauri. Using these measurements, each astronomer then obtained the distances to each of these stars. In each case, the overall result was the same: the stars, even the ones relatively nearby, are hundreds of times more distant than the most remote planets in our solar system. Because of these huge distances, astronomers defined a unit of distance called the light year, which is the distance light can travel during a single year moving at a rate of 300,000 km/sec. One light year is thus, equal to about 10^{13} (or ten trillion) kilometers. The distance to 61 Cygni, Vega, and Alpha Centauri, for example, would be 11, 26, and 4.3 light years, respectively.

A direct consequence of the vastly increased stellar distance scale is that the stars must be energy generators comparable in scale to our own sun in order to be visible. Indeed, we now know that the stars, like the sun, are gaseous thermonuclear infernos. The stars, however, possess a wide range of physical characteristics, many of which can be very unsolar-like. Some stars, such as the so-called white dwarfs are scarcely larger than the earth, while other stars, the red supergiants for example, would engulf a large part of the solar system if placed at its center. Most stars come in pairs, called binary systems, while some, like the sun, are single stars. The bright star, Castor, in the constellation of Gemini, is a system consisting of no less than six stars! There are stars with peculiar chemical compositions and stars which can change their brightness over

*By contrast, the top of Mt. Everest is 8848 meters above sea level.

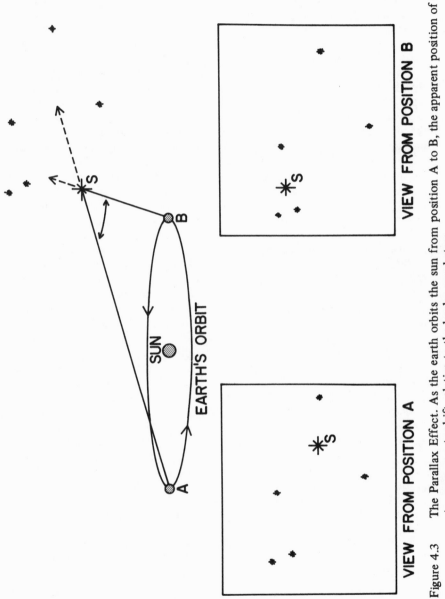

VIEW FROM POSITION B

VIEW FROM POSITION A

Figure 4.3   The Parallax Effect. As the earth orbits the sun from position A to B, the apparent position of the nearby star s appears to shift relative to the background stars.

45

periods of time ranging from a few minutes to several years. To account for this wide variety of "stellar species," astronomers have developed a theory of stellar evolution in which various types of stars are thought of as representing certain stages of a star's "life" or evolutionary cycle. Thus, stars first form out of the interstellar gas and dust and then contract until the temperature and pressure at the center are large enough to initiate the thermonuclear processes by which hydrogen atoms are fused into helium atoms. At this point, the force of gravity tries to compact the star's gaseous material into a small volume while high pressure from the heat of the nuclear fusion reactions attempt to blow the star apart. Remarkably, the net result is a balanced and stable star which can last anywhere from 500 million years to nearly 100 billion years depending on its mass. Ultimately, however, one or the other of these basic forces in a star's life takes charge. In some cases, the thermonuclear processes slip out of control and the star flashes into a nova or supernova explosion in which vast amounts of material are returned into the interstellar medium — material which one day might again be recycled into another generation of stars. On the other hand, the star's own gravity might relentlessly compact the material present into even smaller volumes until finally the star's electromagnetic emanations are stilled by the crushing surface gravity of an asteroid-sized black hole. Such is the modern astrophysicists' portrait of the Greeks' fiery denizens of the star sphere.

It has been the scientific investigations of the medieval "dry exhalations," however, that has produced the most spectacular changes in our overall picture of the physical universe. With his telescopes, Galileo recognized that most of the Milky Way "dry exhalation" was in fact made up of large numbers of faint stars, star "clouds" as it were, which coalesced into a "solid" luminous band of light when viewed with the naked eye. Despite this very important discovery, however, there were still many fixed diffuse patches of light scattered across the sky which could not be resolved into any meaningful detail by the telescopes of the 17th and 18th centuries. By the end of the 18th century, astronomers had begun to catalogue these objects by position so that they would not be mistaken for comets. The most famous of these catalogues were those of Charles Messier (the Messier Catalogue) and Dreyer's 19th century *New General Catalogue of Nebulae and Clusters of Stars* or the "NGC." Even today the most common designations for a diffuse object such as the Great Nebula in Orion are Messier 42 (M 42) and NGC 1976. Since the only requirement for getting into these catalogues is that the object have a diffuse appearance, the NGC and Messier catalogues are veritable grabbags of celestial objects. Many of the diffuse objects lie on or close to the plane of the Milky Way and as such have long been recognized to be a part of the Milky Way system. With the increased resolution of our modern telescopes we have found that these diffuse patches of light include such objects as star clusters, interstellar gas and dust clouds, planetary nebula* and the gaseous debris of stellar explosions. In the course of uncovering the true nature of these objects, our understanding of the Milky Way itself has vastly changed. Starting with Galileo's first telescopic observations, the scientific

*Stars surrounded by a detached shell of gas which gives a disk-like or "planetary" appearance in a telescope.

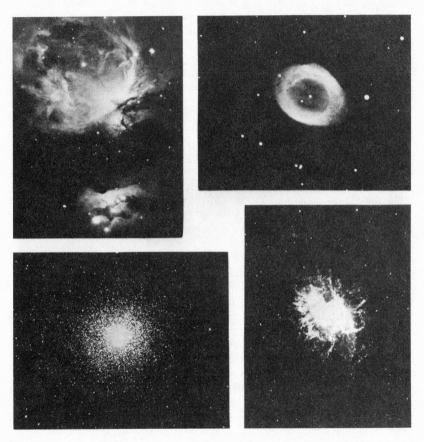

Figure 4.4

Resolution of some of the "Dry Exhalations", top left: the great gaseous nebula M42 in Orion; top right: the planetary "ring" nebula M57 in Lyra; bottom left: the great globular cluster M13 in Hercules; bottom right: the remains of a stellar explosion, the Crab Nebula, M1, in Taurus.

view of the Milky Way has, with ever-increasing observational data, evolved from Herschel's "squished amoeba" model, through the ellipsoidal "Kapteyn" * Universe, to our present view (4.2). The Milky Way which has been uncovered by our 20th century astronomical instrumentation is an awesome pinwheel-shaped star system which is 100,000 light years in diameter and contains roughly 200 billion suns (see Figure 4.5). To appreciate this scale more fully: if the Milky Way were to be reduced in size to one of our Western states such as Colorado or Wyoming, the entire solar system would be correspondingly reduced to the size of a dime!

As large and impressive as the Milky Way turns out to be, however, it is not the last word in the mind-boggling immensity of the universe. In the early decades of this century, astronomers were able to show that large numbers of the diffuse patches were themselves independent star systems (4.3) called galaxies, each of which is comparable in size to the Milky Way. Galaxies by the millions are now known to be strewn across space for hundreds of millions of light years, each one hurtling outward as a part of a rapidly expanding universe. As far as our telescopes can see, we still observe galaxies racing away from us, almost as if they were trying to scurry away from the scrutiny of our instruments.

And what a marvelous array of instruments we have! Starting with Galileo's crude spyglass, the science of astronomy, in response to its need for ever more precise observations and measurements of ever fainter objects, has developed a wide variety of instrumentation. Astronomers now scan the heavens with telescopes capable of detecting a candle's light 60,000 miles away and highly sensitive radio antennae or radio telescopes which can sense radio sources or telemetry signals from a distant space probe which have an apparent power output roughly equal to that of a single Christmas tree light. We can even enhance these views with electronic wizardry at the tailpiece of the telescope. Thus, astronomers have been able to detect star spots on the red supergiant star Betelgeuse in Orion, the interstellar radio "strobe-lights" we call the pulsars, and catastrophic explosions of almost unimaginable size at the centers of some of the distant galaxies.

The universe which has been brought to us courtesy of the scientific method is thus one which is incredibly exciting, complex, and operates within the framework of the same set of scientific laws observed here on the earth. Two objects of unequal mass thus fall to the lunar surface at the same time just as they do here on the earth, and the satellites of Mars and the Gas Giants orbit those planets in the same fashion as the moon orbits the earth. Binary stars light years away do not orbit each other in square or triangular orbits, but rather in ellipses, just as the planets orbit the sun and the moon orbits the earth (see Figure 4.7). The chemical elements that make up the earth are identical to those which compose the moon, planets, distant stars, and galaxies (see Figure 4.8). Of

* Named for the Dutch astronomer J.C. Kapteyn who proposed this model of the Milky Way in 1905.

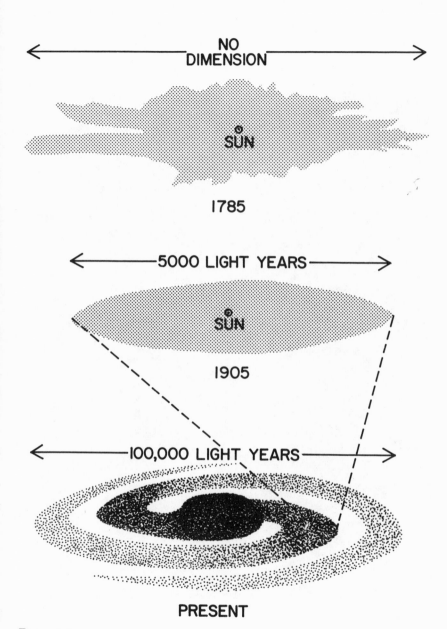

Figure 4.5

The evolution of our view of the Milky Way Galaxy: (top) Herschel's model, (middle) the Kapteyn "Universe", (bottom) the present view.

Figure 4.6

The Hercules cluster of Galaxies. Nearly every image on this photograph is a galaxy millions of light years distant and comparable in size to the Milky Way system.

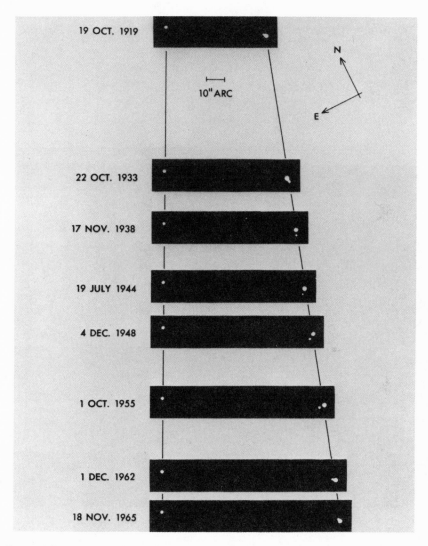

Figure 4.7

"As below, so above-I". The star Kruger 60 B (companion to right hand image) orbits its primary just as the moon orbits the earth, despite the fact that this system is over twelve light years away.

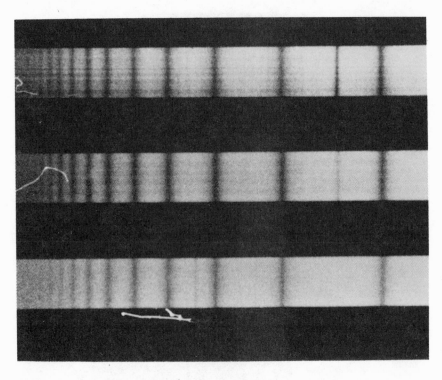

Figure 4.8

"As below, so above-II". Absorption line from the element hydrogen in the light from the stars δ Cas (top), δ Mon, and ρ Aur. The distances to these stars, respectively, are about 110, 220, and 1000 light years.

itself, the principle of universality* presents no particular threat to the astrologer. However, when combined with the known enormity of the universe, the resulting blend, as we shall see, holds some grave consequences for the concept of causal astrology.

Thanks also to the scientific method, the medieval view of the universe has been left in total ruin. Gone are the epicycles, deferents, and the earth's exhalations. Gone also is the earth as the recognized master of the planetary system and the planets as wandering signs of the deities. In short, there is not one aspect of the medieval view of the universe that has not in some degree been altered by the investigations of modern science. Similar statements can also be made for the other sciences as well, including physics, chemistry, geology, and the life sciences. It is against this backdrop of awesome scientific change then that we now examine the progress of astrological thought since the Scientific Revolution.

Let us now return once more to the work of Ptolemy, but this time to the astrologically oriented *Tetrabiblos*. In reading the *Tetrabiblos,* one is almost thunderstruck at the similarities between Ptolemy's astrological principles and those expounded in current astrological texts. For example, Ptolemy tells us that the planet Mercury:

" . . . produces writers, superintends of business, accountants, teachers in the sciences, merchants, and bankers . . . in short all who live by the exercise of literature and by furnishing explanations or interpretations as well as by stipend and salary, or allowance." (4.4)

Over eighteen centuries later, astrologer Manly Hall states that:

"The planets impel those whom they rule into lines of endeavors as follows . . . (Mercury): Accounts, civil engineers, school teachers, letter carriers, journalists, bankers, inventors, orators, booksellers, clerks." (4.5)

As one continues through the *Tetrabiblos,* it is perfectly clear that Ptolemy would have no difficulty whatsoever in establishing himself as a practicing astrologer in the 20th century. On the other hand, a visit by Ptolemy to one of our major astronomical observatories such as Kitt Peak, Mt. Palomar, or the National Radio Astronomy Observatory, would result in a culture shock of literally astronomical proportions. How is it then, that astrology which claims scientific sainthood for itself and its practitioners, has so obviously escaped the upheavals endured by the rest of the modern sciences during the course of history? That they have learned nothing basically new, seemingly does not bother the astrological community in the least. In fact, many astrologers take pride in pointing out that the essence of astrological thought has experienced little in the way of significant changes for centuries. Astrologer Linda Goodman tells us for example that:

*The principle that the laws of nature are invariant with either time or one's position in the universe.

53

"Alone among the sciences, astrology has spanned the centuries and made the journey intact. We shouldn't be surprised that it remains with us, unchanged by time — because astrology is truth — and truth is eternal." (4.6)

Such cool confidence, of course, has long been a resident of the religious realm and more recently, it has become a distinguishing characteristic of the pseudosciences as well. For the true scientist, whose view of the physical world is only as good as the next set of experimental results, this centuries-long immunity to basic alterations in the "scientific" astrological view of nature can only be greeted with the greatest of suspicion.

4.1   Kahn, P., Pompea, S., and Culver, R., *Astron. Qtrly.*, 2, 2, (1978).

4.2   Jaki, S., "The Milky Way: An Elusive Road for Science", Science History Publications, New York, New York (1972).

4.3   Berendzen, R., Hart, R., and Seeley, D., "Man Discovers the Galaxies", Science Publications, New York, New York (1976).

4.4   Ptolemy, C., p. 121, "The Tetrabiblos", Symbols and Signs, North Hollywood, California (1976).

4.5   Hall, M., "Astrological Keywords", p. 118, Littlefield, Adams and Company, Totowa, New Jersey (1975).

4.6   Goodman, L., "Linda Goodman's Sun Signs", p. 475, Bantam Books, Inc., New York, New York (1971).

Chapter 5

The Houses Divided

Happy and sad experiences, success and failure, wealth and poverty, and health and sickness shape the pattern of earthly existence for each one of us. No wonder it has been said that "Life is just one damned thing after another." Moreover these human events in any individual life are intertwined with those in the lives of loved ones, friends, and passing acquaintances. Numerous attempts have been made to explain life's little surprises, but one we are reminded of at this point is the Greek goddess of fortune Tyche rotating her wheel to bring about the great triumphs or disasterous reverses in human affairs. But are these things just a matter of luck or is there some pattern to it all? Astrology claims the heavens reveal the cycle of lifetime occurrences for everyone. At least the celestial signs are clear to the person skilled in their proper interpretation. Thus rather than a spin of the wheel of fortune our life stories are tales told by moving planets and houses from the moment of our birth.

The astrological wheel of worldly affairs usually consists of twelve sectors, or "houses." Most commonly the sectors are numbered in a counterclockwise direction starting from the ascendant, the point which defines the boundary or so-called "cusp," of the first house (see Figure 5.1). Such a grid provides a fixed frame of reference for judging the daily rotation of the celestial sphere, and hence where in a person's affairs a particular planet may impress its influence. The areas of life allegedly represented by the houses are those listed in Table 5.1, but it should be noted that there is always some variation between authors in this matter. As we have listed the spheres of influence, Mars in the first house would indicate an aggressive individual probably with a muscular body. If Mars were found in the eighth house the native might be expected to meet a violent death in war.

The origin of the astrological houses has not been fully traced through

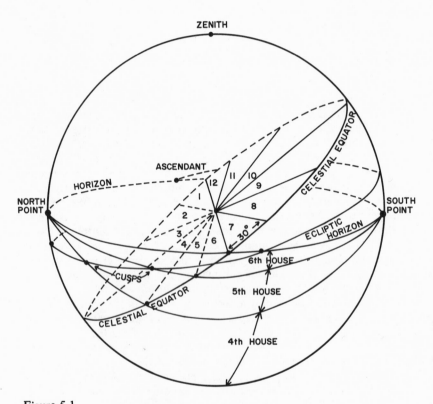

Figure 5.1

The Division of the Astrological Houses by the method of Regiomontanus.

TABLE 5.1   The Astrological Houses

House Number	Sphere of Life Affected (5.4, 5.5)
1	The person, or self; temperament, personality.
2	Possessions and feelings, worldly resources, especially money
3	Short communications, short trips, neighbors, mental interests.
4	The home, family.
5	Creativity; children.
6	Health, work, service.
7	Close relationships of an emotional and business nature; marriage, partnerships.
8	Life force in birth, sex, death; sexual attitudes; legacies.
9	Travel, teaching; larger communications.
10	Profession, matters outside the house, public standing, social status.
11	Ambitions, friends, and acquaintances.
12	Confinement, latent talents; retirement.

history. Again, there may possibly be an astronomical connection here in the attempts of early astronomers to determine time and to measure positions of planets as they moved across the sky. Each planet was considered to dominate an hour of the day, beginning with Saturn. This pattern also produced the names of the days of the week (5.1). Cyril Fagan (5.2) has suggested the earliest chart division was not into twelve sectors but rather into eight, arising from an initial splitting of the solar day and night into four periods, subsequently halved. Since these eight "watch" periods provide a measure of the progress of the sun around the sky in a *clockwise* direction, Fagan has argued that the houses should in fact be ordered in the same way, opposite to the common practice. The areas of influence under Fagan's eight house scheme are just those of the first eight listed in Table 5.1. He has further suggested the origin of these spheres of influence to be in the normal activities of a day beginning with the first watch of sunrise. Thus, the second "watch," in the early hours when the bartering of wares took place, governed money, and the eighth watch or house, a time of slumber, was the death house. Dane Rudhyar has suggested that although this once may have been correct, as society has changed so have man's needs and activities, so the eight house system can no longer apply (5.3). This almost sounds as if it makes sense, but of course it is mere supposition and has utterly no basis in any observational evidence. Nor is there any substantial evidence to support any of the other house systems, as we shall presently see.

The number of house systems so far proposed is estimated to be as large as fifty (some with as many as 24 or 28 separate houses); however, only about twenty apparently have had some use (5.6) and just three or four are widely used by astrologers. They are mostly variations of three basic methods of spreading the houses around the ecliptic. It is of interest to briefly examine how this is accomplished.

The first and perhaps the simplest means of house division is to directly divide the ecliptic according to some specific rule. Dating from antiquity the so-called equal house system sets the cusps of all houses at 30° intervals along the ecliptic starting with the ascendant point and moving counterclockwise. The "M House" system is almost identical but starts at the MC. Other systems may number the houses clockwise, or start with the second house at the ascendant, or the first house 5° above the ascendant, and so on. A method proposed by Porphyry in the 3rd century begins with the ascendant and the MC (10th house cusp) determined in the usual way, then segments the arc length between them into three equal portions (5.7).

A second approach is to divide some great circle other than the ecliptic (the horizon or equator for instance) according to a specified rule and to then project these divisions onto the ecliptic. A couple of examples here might also be useful. In the system developed by Regiomontanus around 1476, the house boundaries are determined essentially by twelve "hour" circles that intersect at the north and south points of the horizon and equally partition the equator. The intersections of these circles and the ecliptic form the cusps of the houses along the ecliptic (Figure 5.1). The system of Campanus, dating from about 1297, sets the house boundaries along 30° segments of the Prime Vertical. The cusps from this system are then determined from the intersection points between the ecliptic and the great circles passing through the north and south points (see Figure 5.2). Additional systems of this sort have been proposed by Morinus, Zariel, and others.

"Trisection of semiarcs" is yet a third way to define the houses in a horoscope. To illustrate we might consider the method of Alcabitius. Having first determined the ascendant, the sidereal time at which this point reaches the MC, or 10th house cusp, is compared to the sidereal time at birth and the difference is divided into three equal parts. These successively added to the birth time define the cusps of the 11th, 12th, and 1st houses, and the remainder are found in an analogous manner. The very commonly used Placidean and Koch "birthplace" systems are members of this family of house division procedures. To understand these systems note that since the ecliptic is in part north of the equator and in part below it, as the celestial sphere rotates a point on the ecliptic will spend different amounts of time above the horizon depending on where it is along the ecliptic. Any particular point traces an arc across the sky known as its diurnal arc. The length at this arc also varies with latitude. In the system devised by the 17th century Placidus de Tito, the 12th house cusp is assumed to be the point on the ecliptic location at sixth of its diurnal arc from the ascendant. The 11th and 10th houses are at the 2/6 and 3/6 diurnal arc places, and so on. The houses "below" the horizon can be found from similar fractions of the "nocturnal" arcs. The "birthplace" system of Dr. Walter Koch defines the house cusps in terms of sixths of the diurnal arc of the MC and nocturnal arc of the IC (Imum Coeli). Although the Koch system has been described as the solution to the problem of house division (5.8), the most commonly used system in the U.S. may still be the Placidean (5.9). This is attributed to the ready availability of house tables in this system rather than it being inherently better.

As is clear from our discussion so far, the positions of the cusps and hence the locations of the houses relative to the signs will vary considerably with the method of house division employed. More seriously, several of the systems fail completely in certain circumstances. For example, the Koch and Placidean systems break down for locations within the arctic and antarctic circles (Figure 5.3) since there are no diurnal arcs for the half of the ecliptic never rising. What uncertain fates must await the millions of people born in these latitudes? And the system of Morinus which utilizes the projections of hour circles originating at the ecliptic poles fails along the tropics of Cancer and Capricorn at certain times (once each day) when these poles are on the horizon. To repeat our comment regarding the popularity of the Koch and Placidean systems, perhaps 90% of the thousands of professional and semiprofessional American astrologers use two systems that do not work over a substantial portion of the earth. We have found one author who suggests such failure should be considered a fatal defect in a house system (5.10).

Even at common latitudes, the disagreement between the various house systems is not trivial. Alan Leo gives an example (5.11) for sidereal time 15 hrs 51 min 15 sec at London, England, part of which we reproduce in Table 5.2 (see also Figure 5.4). It is clear that the house boundary can be located in any one of

59

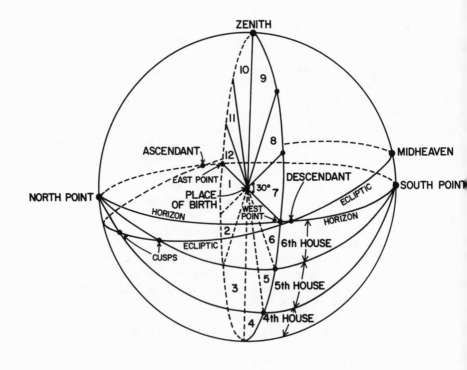

Figure 5.2

The Division of the Astrological Houses by the method of Campanus.

Figure 5.3

The Astrological No-Man's Land. Above 66.5° North latitude, the Placidean and Koch house division systems, those most commonly used by astrologers, fail to generate a viable horoscope, thus depriving 0.3% of the world's population (about 12 million people!) from Kotzebue to Murmansk of their sidereal destinies.

TABLE 5.2   House Boundaries Under Several Methods of House Division
(See Also Fig. 5.4)

House System	12th House Cusp	3rd House Cusp
Equal	27° 15 Sagitarius	27° 15 Pisces
Alcabitius	8° 2 Capricorn	19° 10 Aries
Campanss	22° 18 Sagitarius	18° 7 Taurus
Placidus	5° 57 Capricorn	6° 10 Taurus

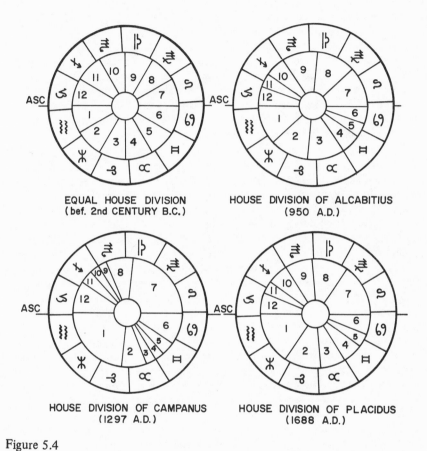

EQUAL HOUSE DIVISION
(bef. 2nd CENTURY B.C.)

HOUSE DIVISION OF ALCABITIUS
(950 A.D.)

HOUSE DIVISION OF CAMPANUS
(1297 A.D.)

HOUSE DIVISION OF PLACIDUS
(1688 A.D.)

Figure 5.4

Comparison of various Astrological House division schemes for the same horoscope.

at least three different signs depending on which system is used. As a further consequence any planet can then fall in completely different houses for one and the same horoscope depending on the system of division used. While one astrologer might place Uranus in the 7th house (native's drive for freedom in a marriage might lead to divorce), another might just as well find it in the 8th house (native is likely to be interested in occult fields), and a third, if he counts the houses in the opposite sense, might have it in the 5th house (unusual romantic involvements, which begin and end suddenly). Cyril Fagan would put Uranus in the 4th house (uncommon home life and unusual family connections). Surely all of the versions cannot be right. Is just one right, or are they all wrong?

Now it seems perfectly obvious that to find out which house system is correct all you have to do is test each one of them experimentally, as science would do, comparing real cases in order to see which one works best, or at least gives a correct description of the native's life the highest percentage of the time. The various systems are well-defined and surely testable without undue difficulty. Nonetheless there have been no adequate tests by astrologers in any organized or thorough manner. What is especially incredible is that the situation is accepted with so little criticism by the astrological community.

Recently a small group of astrologers (5.12), much to their credit, have summarized the statistical evidence regarding house systems showing just how weak such support really is. The studies of Gauquelin (not an astrologer), to be referred to in more detail later, demonstrated in thousands of cases that planets fall in the "wrong" houses or in the "correct" house no more frequently than would be expected purely on the basis of chance (see Table 5.3). The astrologers have essentially ignored these results (5.13). Still other studies (5.14) show results not different from chance, contradictory results, or that the expected effect does not occur. In an analysis of 430 marriage and re-marriage charts (noting Venus transits of the 7th house) for 392 women (5.15) Venus was found not to appear in the horoscopes as expected from traditional astrology. A conclusion can be drawn from the limited evidence: the complicated and sophisticated system of astrological houses does not work.

It has been reported that a 1974 seminar on house division held by leading astrologers degenerated into a shouting match (5.16). In light of the total disagreement over the number of houses, their sequence, the method of division, and their interpretation as well as the lack of empirical foundation to properly evaluate each system, it is little wonder that astrologers often resort to these less esoteric means to resolve their problems. In fact, there may not even be *astrological* justification for the use of house systems (5.17), either by analogy with signs, or by some necessity of interpretation (there are an adequate number of other indicators in the horoscope). A *few* astrologers have therefore abandoned them altogether. No credible evidence exists supporting in any way the predictive value or accuracy of the houses, yet, like the compulsive gambler playing the single numbers at the roulette table, the typical astrologer, in an equally unsuccessful manner, plods along with house systems that are quite often centuries-old relics of the past. If astrologers indeed cared whether or not their procedures worked and if astrology had about it a modicum of the scientific method, the houses would have long ago been razed.

TABLE 5.3   The Number of Times a Given Planet (Sun, Moon, Mercury,
Venus, Mars, Jupiter, Saturn) Fell in a House Traditionally Expected
For a Given Occupation Compared With That Expected Purely
On a Basis of Chance (5.12)

Subjects	House in which Planet is Expected	Average Observed Number	Number Expected by Chance
906 painters	5 (artistic work)	75	76
410 sculptors	5 (artistic work)	32	34
623 criminals	12 (prisons)	55	52
1084 doctors	6 (health, service)	85	90

5.1   Dreyer, J. L. E., "A History of Astronomy from Thales to Kepler", p. 169, Dover Publications, Inc. (1953).

5.2   Fagan, C., "Astrological Origins", p. 161, Llewellyn Publications, St. Paul, Minn (1971).

5.3   Rudhyar, D., "The Astrological Houses", Doubleday and Company, Inc., Garden City, New York (1972).

5.4   Hone, M.E., "The Modern Textbook of Astrology", p. 92, Revised ed., L.N. Fowler and Co., Ltd., London (1973).

5.5   Thompson, D.V., "Chart Your Own Stars", p. 13, Macoy Publishing Co., Richmond, Virginia (1975).

5.6   Dean, G. and Mather, A., "Recent Advances in Natal Astrology", The Astrological Association, Bromley Kent, England (1977).

5.7   Leo, A., "Casting the Horoscope", p. 110, L.N. Fowler and Co., Ltd., London (1969).

5.8   Koch, W.A. and Schaeck, E., Introduction, "Birthplace Tables of Houses", ASI Publishers, Inc., New York, New York (1971).

5.9   Ibid., Ref. 5.1.

5.10   White, H.J., *CAO Times* **3**, No. 3, 46 (1978).

5.11   Ibid., Ref. 5.8, p. 112.

5.12   Ibid., Ref. 5.6, p. 170.

5.13   Ibid., Ref. 5.6, p. 172.

5.14   Ibid., Ref. 5.6, p. 170-173.

5.15   Ibid., Ref. 5.6, p. 173.

5.16   Ibid., Ref. 5.6, p. 165.

5.17   Ibid., Ref. 5.6, p. 170.

Chapter 6

The Age of Aquarius

H.G. Wells once said that human history was becoming more and more a race between education and catastrophe. After nearly eight decades of the Twentieth Century, the race continues as advertised, with the contestants currently being clocked at less than two ICBM flight times apart. In light of the unprecedented difficulties presently facing humanity, it is little wonder that a strong desire exists for some sort of "New Age." Along with all of the New Age promises of this century's "isms," we find that the astrologer, too, is at the ready with an astrological New Age, the Age of Aquarius.

The concept of the Astrological Great Ages is, of course, not new, but astrologers with their customary skill have managed considerable mileage from the "dawning" of the Age of Aquarius, particularly from its popularization a few years ago in the successful Broadway production of "Hair." After all, given the savageries of the current 2000 year old Piscean Age, who can argue with a new era promising that "peace will guide the planets and love will steer the stars"?* Ironically, the Age of Aquarius also recalls one of the oldest aspects of the Gemini Syndrome — the dilemma of precession.

Nearly all of us at one time or other have marvelled at the ability of a top or gyroscope to defy the earth's gravity. The top's secret, of course, lies in its spin. Without it, the top ingloriously tumbles to the ground. But with it, the top becomes an elegant study in rotational motion, with its axis of rotation sweeping out a cone-shaped path as the angular momentum associated with its spin is able to temporarily stave off the ever-present force of gravity. This slow, sweeping motion of the top is referred to as precession and it is of crucial importance in the astrological concept of the Age of Aquarius.

*Lyrics from Hair.

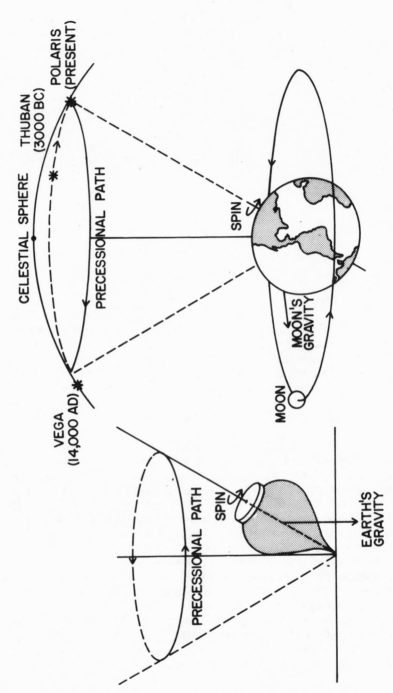

Figure 6.1   The precessional motions of a top and the earth.

As the earth orbits the sun, like the top it also spins about its axis of rotation. And like the spinning top, the spinning earth also defies gravitational forces. The gravity from the sun and moon try to "tip" the earth's rotational axis, while the earth's spin associated angular momentum resists such efforts. As if to enact a sort of classical mechanical compromise, the earth, like the top, precesses along a huge cone-shaped path at a rate of about one precessional cycle per 26,000 years (see Figure 6.1). As a result, the earth's axis of rotation changes its orientation in space, and with it change the so-called celestial poles or places in the sky toward which the ends of the earth's rotational axis seem to point. Currently, for example, the north polar axis of the earth very nearly points to the bright star Polaris in the constellation of Ursa Minor. It was not always so, however. Some 5000 years ago it pointed toward the bright star Thuban in the constellation of Draco, and 12,000 years from now, the earth's inhabitants will enjoy the brilliant Vega as their "pole star." If we journey along the celestial sphere south from the pole star, we find that another key point of the heavens, the vernal equinox point, also appears to move relative to the background stars as a result of precession (see Figure 6.2). To be sure, the 1/72 degree/year rate at which the equinox points ply along the ecliptic is not overwhelmingly large, but down through the ages, as the years have added up into centuries and millennia, the precession of the equinoxes has begun to pose several interesting questions for the astrologer.

First of all, the precession of the equinoxes has brought about the disconcerting and confusing result that the astrological signs of the zodiac, which are usually measured relative to the vernal equinox point, have become displaced with respect to the constellations for which they are named. For example, the dates in which the sun is said to be located in the astrological sign of Scorpio extend from October 24 to November 22 (6.1). If we plot the sun's actual position on those dates, however, we find that it is not in the constellation of Scorpius, but is to the west in Virgo (see Figure 6.3). On November 23, as the sun enters the astrological sign of Sagittarius, we find that it is now located in the constellation of Libra, two astronomical constellations to the west. This discrepancy between astrological signs and astronomical constellations forces the astrologer into a difficult position. Is an individual born on November 5, for example, a Libra or a Scorpio? It is not a trivial question, since the astrological properties of these two regions are vastly different not only for the sun-sign, but for the locations of all of the other astrologically significant items such as the moon, planets, ascendant, etc. as well. A common error made by critics of astrology at this point is to *equate* astronomical constellations with astrological signs, make some clever remark about sexy Scorpios really being balanced Libras, and then exit the scene. Actually, the effect of the precession of the equinoxes has been known to astrologers at least since the time of Hipparchus ($\sim$ 150 BC) and perhaps since the time of the Babylonian Kidinnu of Sippar roughly two centuries earlier (6.2, 6.3). The real problem posed for astrologers by precession is whether one wishes to tie one's horoscopic fortunes to a zodiac that moves with the vernal equinox point, ("tropical" zodiac) or one that remains anchored to the set of "fixed" astronomical constellations ("sidereal" zodiac).

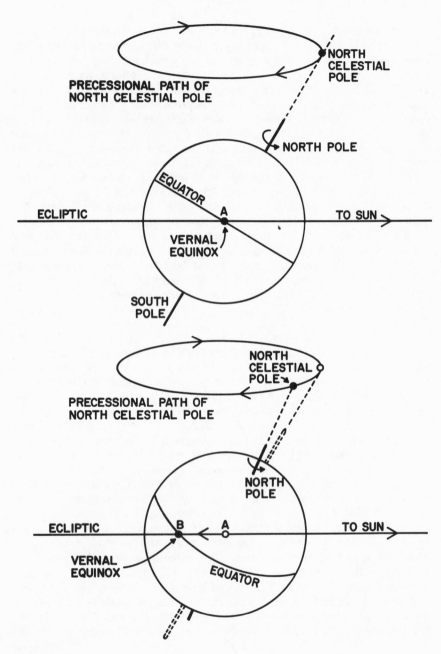

Figure 6.2

The motion of the vernal equinox point due to precession.

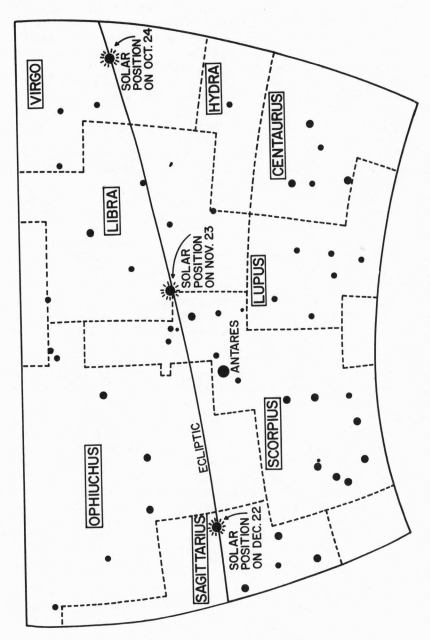

Figure 6.3   The sun's apparent path among the fixed stars from October 24 to December 22.

Most modern astrologers opt for the tropical zodiac in which the zero point of the zodiac or "First Point of Aries" is marked by the vernal equinox point. The "tropical" astrologers thus, contend that while there was a correspondence between the names of the astrological signs and their respective constellation namesakes at the time of Ptolemy's *Tetrabiblos,* this correspondence no longer exists because of the effects of precession. For the tropical astrologer, then, the astrological signs are essentially defined on the basis of the astronomical seasons, with Aries, Taurus, and Gemini being the springtime astrological signs and so on. The system has the added advantage that the vernal equinox point is the zero point for several astronomical coordinate systems, and as such, has its movements among the stars precisely charted by positional astronomers. Thus, all is well with the tropical astrologer, except for one small problem — The Age of Aquarius.

In the course of its leisurely 26,000 year circuit about the heavens, the vernal equinox point naturally passes through each of the astronomical constellations that line the ecliptic. The temptation to somehow incorporate this effect into the astrological ediface has proven to be too much for astrologers, and they have thus developed the concept of the "Great Astrological Ages." Basically the idea is that human history can be divided into intervals of roughly 2160 years, or 1/12 of the vernal equinox's "Great Year." The characteristics of each age then correspond to those of the astrological sign in which the vernal equinox point is located (6.4). The current Age of Pisces, for example, can be thought of as having for its terrestrial correspondence the Christian fish, the symbol of humanity's most dominant religious movement during the past two millennia (see Table 6.1). The dawning of the much-storied Age of Aquarius will thus occur as the vernal equinox point crosses the Pisces/Aquarius border into Aquarius.

But for the tropical astrologer the nectar of the Age of Aquarius is a self-made lure into a celestial Venus fly-trap. Having fled the sidereal safety of the astronomical constellations, the tropical astrologer is now faced with the fact that in the tropical zodiac, as the vernal equinox goes, so goes the sign of Aquarius. Regardless of how long the vernal equinox struggles along the ecliptic, every time the tropical astrologer looks up, he or she finds that not a single arcsecond has been trimmed off the 30 degree span between the First Point of Aries and the Pisces/Aquarius cusp. Thus, for the tropical astrologer using a tropical zodiac, there can never be a dawning of the Age of Aquarius. The only way that the tropical astrologer can arrange for the needed celestial backdrop for humanity's grand entrance into the Age of Aquarius is to provide Aquarian characteristics to the *astronomical constellation* of Aquarius. Only then does a vernal equinox crossing of the Pisces/Aquarius border hold any signficance. If the tropical astrologer is willing to do this, however, then we must immediately return to the zodiacal "Who's Who" kinds of questions discussed earlier in the chapter.

One obvious way out of the difficulty is offered by the sidereal astrologers whom West and Toonder (6.5) refer to as "a small but noisy contingent" in the astrology world. Unlike their tropical counterparts, however, the sidereal

72

TABLE 6.1   The Astrological Ages

AGE	APPROXIMATE DATES	CHARACTERISTICS
Leo	10,000 BC – 8,000 BC	Leonian creativity is liberated as Leo's ruling planet, the sun, rolls back the Ice Age glaciers
Cancer	8,000 BC – 6,000 BC	Under the strong family and fertility influences of Cancer and its ruling planet, the Moon, human beings emerge from caves to establish settled farming and fixed dwellings.
Gemini	6,000 BC – 4,000 BC	"Lively intellectual, mobile and communicative influence of Gemini" as evidenced by the first widespread use of the wheel, writing, and collective scholarship.
Taurus	4,000 BC – 2,000 BC	Aldebaran, the "eye of the bull," marks the vernal equinox point and myths and legends involving bulls abound as do bull-god cults.
Aries	2,000 BC – 0 BC	The ram replaces the bull as the sign of worship; sacrifice of sheep is common; war-like tendencies of Age abetted by Mars, Aries' ruling planet. Astrological tropical and sidereal zodiacs coincide.
Pisces	0 AD – 2,000 AD	Correspondence between Pisces and the Christian symbol of the fish summarizes the chief movement of the Age.
Aquarius	2,000 AD – 4,000 AD	Unprecedented technological advance as a result of the influence of Uranus, the planet of surprise and discovery. Aquarian freedom, understanding, and brotherhood to give science a "human face."

astrologers can effectively have their Age of Aquarius cake and eat it too. Since the sidereal zodiac has not been moving, there is no discrepancy between the astrological sign of Aquarius and its constellation namesake. Thus, for the sidereal astrologer, the constellations and astrological signs are one and the same. Hence, the dilemma posed above for the tropical astrologers is neatly bypassed. Despite the fact that sidereal astrologers at least possess internal consistency, they do not go untouched by the Age of Aquarius.

As we have already noted, the dawning of the Age of Aquarius is defined as the time at which the vernal equinox point precesses across the Pisces/Aquarius border. The difficulty arises in locating that border relative to the fixed stars. For the astronomer the problem is trivial. Since there is no astronomically magical or mystical significance tied to locating constellation borders, astronomers have simply defined where they are in the sky. Thus, the current boundaries of the constellation of Aquarius, like those of all the other constellations, were formally agreed to in 1928 by the International Astronomical Union. While these constellation boundaries more or less correspond to those on most earlier star maps, there has been considerable variation in such borders on the star maps constructed down through history (see Figure 6.5). As a result of this one point alone, there is considerable disagreement as to when the vernal equinox's storied border crossing will occur.

In raising the question of the astrological effects of constellation boundaries, however, the sidereal astrologers plow themselves right into a sidereal version of the Qattara Depression. For example, not all of the signs of the sidereal zodiac intercept exactly 30 degrees of ecliptic. The boundaries of Scorpius include only 7 degrees of the ecliptic while Virgo greedily grabs off some 44 degrees (see Table 6.2). Moreover, on the basis of constellation boundaries, the ecliptic has two celestial interlopers in Ophiuchus and Cetus. This fact prompted sidereal astrologer Steven Schmidt to write his book *Astrology 14* (see Chapter 7) in which an attempt is made to weave these two "scientifically confirmed" astrological signs into the overall astrological fabric.

Nor is there relief for the sidereal astrologer attempting a 90° turn off the ecliptic to the north or south. If the astronomically defined constellation boundaries are to be regarded as astrologically significant, then once a departure is made from the ecliptic, it becomes possible for the moon and planets to wander into no less than the 29 additional constellations listed in Table 6.3! To illustrate, we have plotted in Figure 6.6 Pluto's 1968-1973 journey across Coma Bernices and Venus' 1975 loop through Sextans. One might argue that some of the constellations in Table 6.3 such as Microscopium and Sextans are relatively new on the celestial scene, hence do not possess astrological significance. Most of the constellations in Table 6.3, however, possess historical credentials every bit as impressive as any from the zodiac. Scorpius and Orion, for example, are in fact usually tied together in mythology, with Orion being given a place in the stars for his many daring deeds, and Scorpius for handing Orion the only loss of his career.

In short, the sidereal astrologers' preoccupation with the actual astronomical constellation zones raises the much more fundamental issue of an astrologically imprintable space, an issue which we will consider in more detail in Chapter 8.

74

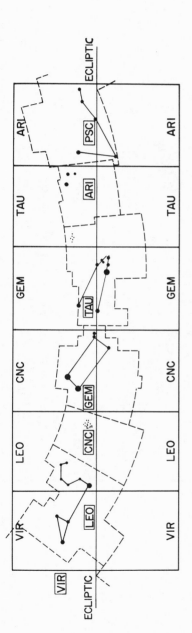

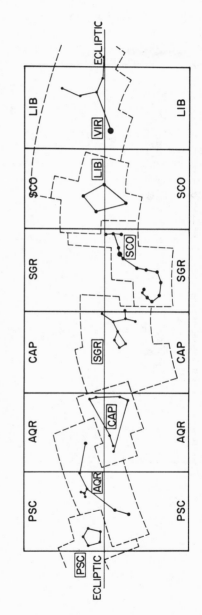

Figure 6.4　The Tropical Zodiac (solid rectangles) and the modern constellation boundaries (dashed lines). These constellation boundaries roughly correspond to the boundaries of the various versions of the sidereal zodiac. Because of the effects of precession, the tropical astrological signs have slipped westward by nearly one complete sign relative to the sidereal signs.

75

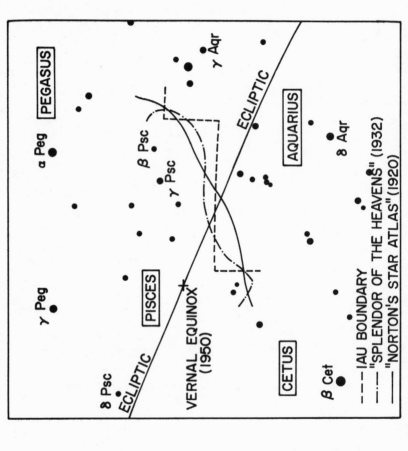

Figure 6.5   The Pisces/Aquarius border according to the International Astronomical Union, Phillips "Splendour of the Heavens" (6.6), and Norton's Star Atlas for 1920 (6.7). The starting date of the Age of Aquarius for these

TABLE 6.2. The Ecliptic Constellations

CONSTELLATION	ECLIPTIC INTERVAL* (DEGREES)
Aries	24.5
Taurus	36.5
Gemini	28.5
Cancer	20.0
Leo	35.5
Virgo	44.0
Libra	23.0
Scorpius+	7.0
Ophiuchus	18.0
Sagittarius	33.5
Capricornus‡	29.5
Aquarius	22.5
Pisces	37.5
Cetus**	0.0

*International Astronomical Union (IAU) boundaries

**The northwest corner of Cetus just touches the ecliptic at a point approximately 7 degrees east of the 1950.0 location of the vernal equinox point.

+Corresponds to the astrological sign of Scorpio

‡Corresponds to the astrological sign of Capricorn

TABLE 6.3 Non-Zodiacal Constellations Through Which One or More Astrological Planets Can Pass*

CONSTELLATION		CHRONOLOGICAL ORIGINS
Andromeda	chained lady	Earlier than 6th Century BC
Aquila	eagle	Earlier than 6th Century BC
Auriga	charioteer	Earlier than 6th Century BC
Bootes	herdsman	Earlier than 6th Century BC
Canis Minor	lesser dog	Earlier than 150 BC
Centaurus	centaur	Earlier than 6th Century BC
Cetus	sea-monster	Earlier than 6th Century BC
Coma Bernices	Bernice's hair	1437 (Ulugh Beg)
Corona Australis	southern crown	Earlier than 150 BC
Corvus	crow	Earlier than 6th Century BC
Crater	cup	Earlier than 150 BC
Eridanus	river	Earlier than 6th Century BC
Hydra	sea-serpent	Earlier than 6th Century BC
Leo Minor	lesser lion	1690 (Hevelius)
Lepus	hare	Earlier than 270 BC
Lupus	wolf	Earlier than 6th Century BC
Lynx	lynx	1690 (Hevelius)
Microscopium	microscope	1763 (Lacaille)
Monoceros	unicorn	1624 (Bartsch)
Ophiuchus	serpent-bearer	Earlier than 6th Century BC
Orion	hunter	Earlier than 6th Century BC
Perseus	hero	Earlier than 270 BC
Pegasus	winged horse	Earlier than 6th Century BC
Piscis Austrinus	southern fish	Earlier than 6th Century BC
Sculptor	sculptor	1763 (Lacaille)
Scutum	shield	1690 (Hevelius)
Serpens	serpent	Earlier than 150 BC
Sextans	sextant	1690 (Havelius)
Triangulum	triangle	Earlier than 270 BC

*International Astronomical Union (IAU) boundaries

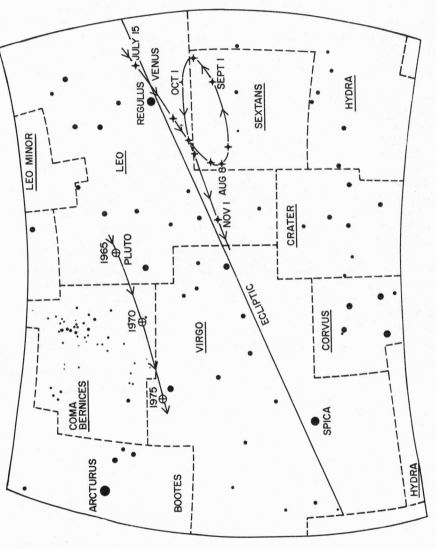

Figure 6.6   The motion of Venus through Sextans in 1975 and Pluto through Coma Bernices in the late 1960's and early 1970's.

As if the border trouble weren't enough for Aquarian Age astrologers, there are also astrologers who feel that the Age of Aquarius should be dated not on the basis of the meanderings of the vernal equinox point, but on historical events themselves. One of the most popular beginning dates in this regard is the March 13, 1781 date of Herschel's discovery of Uranus, the planet of "discovery and surprise" and the "true" ruler of the astrological sign of Aquarius (see Chapter 11).

Still other astrologers claim that one cannot assign a precise start and end to an era of history. These individuals argue that, as in any change of historical eras, there exists a sort of intermediate transition period in which one finds a mixture of the characteristics of each age. It is thus no more appropriate to set an exact date for the beginning of the Age of Aqaurius than it would be for the beginning of any other historical era such as the Western European Renaissance.

The net result that emerges from all of this is a rather handsome spread of astrological "dawnings" of the Age of Aquarius, ranging from 1781 to 2740 (Table 6.4). These data are even more impressive if one recalls that the entire span of the Aquarian Age itself is only about 2160 years! The Table 6.4 results, however, are only symptomatic of astrological difficulties of a much more fundamental nature. Once again, as in the problem of astrological house division, we find the astrologer confronted with competing models that offer us a relatively simple choice, this time the choice between sidereal and tropical zodiac. Once again, we find that a clear-cut choice between these models is not made, despite the fundamentally important astrological overtones that are involved. Instead, we find that the sidereal and tropical astrologers argue abstractly for their models, with both advancing an impressive array of conflicting claims and properties for their respective versions of the astrological zodiac. Unfortunately, such has been the case in astrology almost since the discovery of precession, and such will continue to be the case until astrologers are willing to submit the conflict to empirical arbitration. By contrast, it was just such arbitration that allowed 17th century scientists to decide between the heliocentric and geocentric models of the planetary system. In short, for the third time in three chapters we have found the astrologers either unwilling or unable to set their empirical affairs in order.

TABLE 6.4   The Dawning of the Age of Aquarius

STARTING YEAR	REFERENCE/COMMENT
2614	Year vernal equinox point crosses IAU Psc/Aqr border
1781	Gleadow (6.9)
1844	Rudhyar (6.10)
1900	Leek (6.11)
1904	Righter (6.12)
1962	Stearn (6.13)
1983	Wilson-Ludlam (6.14)
2000	Parker & Parker (6.4)
2160	Oken (6.15)
2217	Gatti (6.16)
2375	Gleadow (6.9)
2376	Fagan (6.17)
2740	Hone (6.18)

6.1   Parker, D. and Parker, J., "The Compleat Astrologer", p. 121, McGraw-Hill Book Company, New York (1971).

6.2   Dampier, W., "A History of Science", p. 45, Cambridge University Press, London (1971).

6.3   Jones, W. and Heath, Sir T.L., "Hellenistic Science and Mathematics", p. 241, Cambridge Ancient History, Vol. VII (1927).

6.4   Ibid., Reference 7.1, pp. 44-45.

6.5   West, J. and Toonder, J., "The Case for Astrology", p. 142, Penguin Books, Inc., Baltimore, Maryland (1973).

6.6   Phillips, T., "Splendour of the Heavens", p. 905, Robert M. McBride and Company, New York (1932).

6.7   Norton, A., Maps 3 & 4, "A Star Atlas and Telescopic Handbook", Gall and Inglis, London (1921).

6.8   Jobes, G. and Jobes, J., "Outer Space: Myths, Name Meanings, Calendars", pp. 279-282, Scarecrow Press Inc., New York (1964).

6.9   Gleadow, R., "The Origin of the Zodiac", p. 55, Atheneum Press, New York (1969).

6.10   Rudhyar, D., *Horoscope*, April 1978, p. 99.

6.11   Leek, S., "Moon Signs", p. 19, Berkley Publishing Corporation, New York (1977).

6.12   Righter, Carroll, *Time*, March 21, 1969, p. 47.

6.13   Stern, J., "A Time for Astrology", p. 19, Signet Books, New York (1972).

6.14   Wilson-Ludlam, M., "The Power Trio", McCoy Publishing Co., Richmond, Virginia (1977).

6.15   Oken, A., "Astrology: Evolution and Revolution", p. 22, Bantam Books, Inc. New York (1976).

6.16   Gatti, A., *Astrology*, December 1977, p. 45.

6.17   Fagan, C., "Astrological Origins", p. 100, Llewelyn Publications, St. Paul, Minnesota (1973).

6.18   Hone, M., "The Modern Text-Book of Astrology", p. 279, L.N. Fowler and Company, Ltd., London (1973).

Chapter 7
Donning the Empirical Trappings

If, as the wisdom of the ages suggests, imitation is indeed the sincerest form of flattery, then one of the finest tributes ever paid to modern science comes to us directly from the realm of the astrologers. For in its struggles to regain its position of preeminence in Western society, which was lost in the ravages of the Scientific Revolution, the astrological chameleon has sought, with considerable success, to take on the exterior tones and colorations of modern science.

The computer, for example, is one of the items in our society that stands almost synonymously with 20th century science and technology, and, sure enough, we find that along with telephone bills, course grades, and ICBM trajectories, the computer has also been tapped to crank out "modern" horoscopes. Typical of the promise held out to would-be clients by computerized horoscopes is the following segment from an advertisement in "Astrology" magazine:

> "You get precise calculations that only a computer can provide without error. From exact time, date, and place of birth, our IBM 370-155 sorts out 24,000,000 bits of astrological information." (7.1)

And for the astrologically minded who do not have access to a computer such as the IBM 370-155, Agrey-Thatcher expounds on the use of the pocket electronic calculator in astrology (7.2).

One of the more damaging trappings of modern science in the minds of the public at large has to be its ever-increasing reliance on scientific jargon. Unfortunately, in the course of communicating complicated experimental results and difficult theoretical concepts it has often become necessary to use language that is virtually incomprehensible to the non-scientists. For example, in a 1975 paper, we ourselves wrote:

" . . . It is a composite of an F star having an MK type of F3 IV and an S-star, with the only evidence of the S-star at this dispersion being three very weak and indistinct bands at $\lambda\lambda 4737$, 4761, and 4804 presumably due respectively, to ZrO, TiO, and TiO . . . Photoelectric UBV observations of the T Sgr system were obtained . . . with a cooled IP21 and DC charge-integration system with KPNO UBV filter set No. 2 and 10s integration time . . . Suitable standards . . . were observed to affect the transformation from the instrumental to the UBV system; mean extinction coefficients and standard KPNO programs were used in the reductions . . . " (7.3)

The problem, of course, is that to an uncomprehending public jargon is jargon, whether it comes from the astronomy or the astrology side of the aisle. And it indeed comes from the astrology side of the aisle, whether it be while explaining the principles underlying the zodiac:

"But polarity is incorporated *within** triplicity. Three is not merely two plus one – though it sounds illogical put that way. In esoteric terms, the descent of unity into multiplicity is sometimes expressed as: one becomes two *and** three simultaneously." (7.4)

or just a good old-fashioned horoscopic interpretation:

"Pluto, co-ruler of your ascendant, is found in your eighth house of death, sexual potential, partnership assets, in Cancer; it is sextile your Sun, so your project has a base for success. Your ninth house of higher mind, religion, education holds Neptune in the 29th degree of Leo sextile Mercury and Venus in Libra in your eleventh house of hopes and wishes. This vibration is an opportunity one . . . Your sixth house of work and health holds Jupiter retrograde Taurus, adverse to the Moon but widely sextile Pluto . . . " (7.5)

Clearly such items have only superficial similarities. It is when one examines the astrological attempts to don the various articles of the wardrobe of the scientific method, however, that the truly distinctive fissures appear between scientists and astrologer.

As we have already seen, the modern scientific method consists of four basic components including experimentation, recognition of laws or behavior patterns in the experimental results, describing the entire collections of laws with an overall theory, and predictive testing of the theory. Perhaps the most elegant of these components is experimentation. It is the linchpin of the scientific method. It permits the development of scientific laws and the testing of theories. It is the bridge between the "real world" and our scientific representations of that world, and it is the final arbitrator of any and all scientific disputes. Thus, it is with no small amount of consternation that the modern scientist greets the techniques by which empirical data are usually generated by the astrologer.

Throughout the natural sciences, for example, there exists an array of basic physical constants of nature, such as the speed of light, the mass of a proton, the

*Authors' italics

84

universal gravitational constant, etc. which constitute an important aspect of our description of the physical universe. Each of these constants has a precise value which has been determined through carefully planned and executed experiments. Moreover, each of these values has associated with it an observational uncertainty, which is essentially a rating of the precision with which the determination was carried out.

In a similar vein, the astrologers, too, have a set of "astrological physical constants" which constitute an important aspect of the astrological universe; some of which we have listed in Table 7.1. Like the constants of the natural scientists, they, too, have both specific values and associated uncertainties.

Perhaps the most familiar of the astrological constants in Table 7.1 is the 30° length of an astrological sign. The value of 30° is, of course, a direct consequence of the astrologers' adoption of a twelve sign zodiac. Interestingly, various cultures throughout history have, at one time or other, employed at least ten different schemes of zodiacal division ranging from an early Euphratean six-sign zodiac to the 28 signs in one of the Chinese versions (see Table 7.2). The debate and rationale behind the general adoption of the familar 12-sign zodiac, and its 30° astrological signs is largely buried in antiquity. Most authors on the subject, however, tend to support the view that the current system took its inspiration from the fact that twelve lunations occur during the course of the sun's annual trek through the zodiacal signs. Regardless of the details of its early background, however, the twelve sign zodiac which swept out of the plains of Mesopotamia, has down through the centuries won the hearts of astrologers the world over. In fact, the current fervor with which the astrologers regard the twelve sign zodiac rivals that of the medieval Church for her cherished Ptolemaic planetary system.

In 1970, for example, astrologer Steven Schmidt wrote a book entitled *Astrologer 14* which was balleyhooed as "the most exciting discovery in astrology in two thousand years!" Basically Schmidt claimed that since the sun passes through the two astronomical constellations of Cetus and Ophiuchus, these star groups should be a part of the system of astrological zodiacal signs as well. In formulating the astrological character traits for each of his fourteen zodiacal signs, Schmidt described his method thusly:

" . . . I collected people — people born under all fourteen signs — and examined their character traits. I used persons with whom I am personally acquainted and also famous people whose personalities are well-known* (statesmen, movie stars, scientists, artists, etc.). I looked for traits in common by persons born under the same sun sign. Always, of course, I kept in mind the wide diversity made possible by differences of heredity and environment." (7.13)

At no time throughout this 144 page presentation does Schmidt put forth any quantitative or statistical data or in any way present any empirical justification for his value of 25.7° for the angular length of an astrological sign in a fourteen sign zodiac. Despite this fact, however, his hope was that the astrological

*Included in this list of "well-known" personalities are Presidents Rutherford Hayes, Chester Arthur, and Millard Fillmore.

TABLE 7.1   A Partial List of Fundamental Astrological Constants

CONSTANT	VALUE	ASTROLOGICAL UNCERTAINTY
Angular Length of Astrological Sign	30°	1°
Angle of Conjunction	0°	8°-12°
Angle of Sextile	60°	4°-10°
Angle of Square	90°	8°-10°
Angle of Trine	120°	8°-10°
Angle of Opposition	180°	8°-12°

TABLE 7.2   The Zodiac In History

CULTURE	NUMBER OF SIGNS	ANGULAR LENGTH OF SINGLE SIGN (Degrees)	REFERENCE
Early Euphratean	6	60	7.6
Early Euphratean	8	45	7.7
Early Euphratean	10	36	7.8
Babylonian	11	32.7	7.9
Many Cultures	12	30	7.10
Modern Western	14	25.7	7.11
Babylonian	18	20	7.9
Toltec	20	18	7.12
Chinese	24	15	7.10
Chinese	28	12.8	7.10

community would "have the courage and foresight to cast off outmoded 'traditional' data". At this point, the traditional astrologers could have brushed aside Schmidt's contentions with a single, simple reference to some classical experiment or series of experiments in astrology which yield the Table 7.1 value of 30° ± 1° for the angular length of a single astrological sign. The sad fact of the matter is that no such reference exists, and without it, the astrologers' traditional 30° value rests on ground no firmer than Schmidt's value of 25.7°. The dispute between Schmidt and the astrological community over the issue of the number of signs in the zodiac thus provides us with not only a marvelous present-day example of medieval science in action, but also a gentle reminder of the need for an empirical method to scientifically settle such disputes.

Similar comments can be made for the astrological aspects. To begin with, the 360° circular ecliptic can be evenly divided by the following integers: 2, 3, 4, 5, 6, 8, 9, 10, 12, 15, 18, 20, 24, 30, 36, 40, 45, 60, 72, 90, 120, and 180. Clearly the division of the circle by the integers 2, 3, 4, and 6 generate the "primary" or "powerful" aspects of opposition, trine, square, and sextile, which represent angular separations, respectively, of 180°, 120°, 90°, and 60°.* This leaves us, then, with 18 other integers which also divide evenly into 360°, but have been excluded from the list of primary aspects. A few of these aspects including the quintile (separation = 72°), semi-square (separation = 45°), and semi-sextile (separation = 30°) are included in the scheme, but are relegated to a "moderate", "weak", or "secondary" status. Complicating the matter even further is the fact that one can also generate astrologically significant aspects from rational fractions of 360°. Thus the sesquiquadrate (separation = 135°), the quinounx (separation = 150°), and the biquintile (separation = 144°) are included by some astrologers as secondary aspects and represent, respectively, 3/8, 5/12, and 2/5 of a 360° ecliptic. In effect there are a fair number of possible "nice" aspects which have either been relegated to second class status or have been turned away completely empty-handed. At this point we would expect to be presented with the empirical basis upon which the astrologers' aspect choices are made, but alas, aside from the very recent work of the "harmonic" astrologers such as John Addey (to whom we will return shortly) the basis for the astrologers' aspects once more lies deeply buried in the "experience" or "wisdom" of the ages. One can contrast this situation with that of the modern scientist, who, with a minimal effort can locate one or more references to experiments in which the value of a certain physical constant in nature has been determined. Jenkins and White (7.14), for example, discuss no fewer than ten different experimental techniques for measuring the speed of light!

Perhaps the most bothersome aspect of the astrological constants, however, is the fact that each and every one of them comes out to be a "nice" number. In fact there seems to be an effort, either consciously or subconsciously, on the part of the astrological methodology to hold on to the "nice" numbers and discard the "not-so-nice" ones. Thus, Mr. Schmidt's 14 sign zodiac and the

*The conjunction (separation of 0°) can be thought of as being generated by dividing the 360° circle by the integer 1.

88

ancient "septile" have been discarded from horoscopic calculations as much for the fact that 360 ÷ 14 and 360 ÷ 7 don't come out evenly as from any other consideration.

The experienced experimental scientist can only watch such antics with great amusement, for seldom, if ever, do the physical constants that he or she measures even come close to "nice" values. To illustrate the point, one can draw a square, which has a nice even value such as 10 cm. for each of its sides. If one then measures the length of the square's diagonal, one doesn't get a "nice" value but a value equal to the side of the square times the $\sqrt{2}$ or 1.414121 . . . Physical constants in nature can have even nastier values. In the metric system, the universal constant of gravitation, G, for example, has a value of

$$.00000006668 \text{ dynes} \cdot cm^2/gram^2$$

The speed of light, c, checks out at 299,793 km/sec and the mean earth-sun distance, or astronomical unit, has a value of 149,597,893 km. In fact, so diverse are the values for the quantities dealt with in science that they are usually expressed in the convenient "scientific" notation in which a given value is expressed as the product of a number between one and ten raised to the appropriate power.*

Closely related to the experimental results generated by the scientist are the uncertainties associated with such measurements. These uncertainties arise from the fact there is an ultimate limit on the accuracy with which a given measurement can be made with a given piece of equipment. Thus, one could use a simple ruler to measure the width of the top of a desk accurately to about a millimeter or so, but that same ruler could not be employed if a more accurate value were desired. These uncertainties in scientific measurements quite often play a critical role in the conduct of the scientific method. Suppose, for example, that two scientific theories predict the two curves shown in Figure 7.1, and suppose that in attempting to test which theory is valid, the observed set of points is experimentally determined. If the uncertainties associated with the measurements are high as they are in Figure 7.1(a), then *within the uncertainties of the measurements* the two theories are indistinguishable. If, on the other hand, the uncertainties associated with the measurements are small, as they are in Figure 7.1(b), then only theory A is consistent with the experimental results and theory B must be discarded. Because of this potentially crucial role, the uncertainty associated with an observation or measurement in science is always formally defined in terms of a statistically meaningful concept, such as the so-called standard deviation.**

The astrological trappings corresponding to the scientists' standard deviations are the orbs. Basically the orb is an interval of angle centered on a particular aspect over which the effects of that aspect are felt. Thus, if the orb associated with the square aspect is 4°, then the aspect will exert influence in the chart for

*In scientific notation, for example, $G = 6.668 \times 10^{-8}$ dyne·$cm^2/gram^2$ and c = 2.99793×10^5 km/sec.

**A concept which can be found in any elementary statistics text.

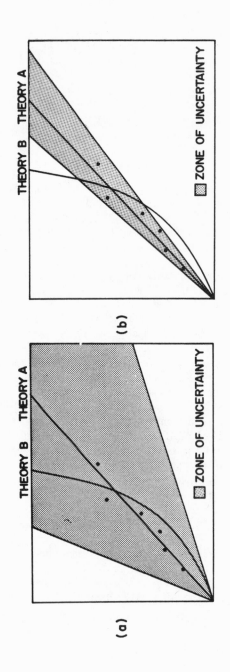

Figure 7.1   The Effects of Experimental Uncertainty on the Scientific Process. Because of the large zone of uncertainty surrounding the experimental points in graph (a), these experimental results cannot distinguish the curve predicted by Theory A from that predicted by Theory B. The same experimental points in graph (b) with their smaller zone of uncertainty are able to empirically eliminate Theory B.

separations of the two objects involved ranging from 86° to 94°. The values of the orbs, then, are of paramount importance, for they essentially determine whether or not one experiences the effects of a given aspect. It is with great surprise, then, that one finds a somewhat blasé treatment of them in the astrological literature. In Table 7.3, for example, we have listed the principal aspects and their associated orbs as quoted by various astrological sources. Incredibly, this crucial quantity not only varies by as much as a factor of two in some cases, but also, depending on the author, can have an uncertainty of its own! In attempting to ferret out the methods by which the orb values are determined in astrology, once more we find that they are conveniently buried in astrological history and experience. Sandra Shulman, however, gives a partial clue when she writes in her definition or orb

"If the aspects had to be absolutely accurate they would not occur so often — hence the orb." (7.20).

To illustrate the impact of the orbs on a typical horoscope, we have listed in Table 7.4 all of the principle aspects for the horoscope presented earlier in Chapter 1, along with the orbs assumed by the astrologer involved. We have also listed the principal aspects for angular separations within 1° of the exact aspect. In the case of the exact aspect, the aspect devastation is virtually complete. Of twenty-three principle aspects, only two are aligned within 1° of their precise values. Clearly, then, the orbs serve the astrologer as impressive generators of "aspect action" in the horoscope. We will leave it to the reader to decide whether this result is inadvertent or by design.

Throughout history the backbone of astrological thought has always been found in the "signs in the stars" or the so-called astrological correspondences. Like their counterparts in the physical laws of modern science, the astrological correspondences seek to establish patterns of behavior in the physical world from which we can gain better insight and understanding as to the workings of that physical world. Unfortunately, in attempting to establish such correspondences, astrologers all too often have not only taken liberties with the methods of empirical science, but assault our very credulity as well. A particularly exquisite example comes to us from the November 1974 issue of "Horoscope" magazine. In an article entitled "Comets: Modifiers of Human and Historical Trends," astrologer M.A. Smollin writes:

"What the coming of a prodigious comet presages, thus will be of complicated and lasting consequence. It was so with the famous comet of 1774 with its spectacular six tails. It, too, filled mankind with expectation. It brought to many fear, suspense as to what might await them in the days ahead." (7.21)

Mr. Smollin then lists an impressive number of events in 1774 that he associates with this comet, including the First Continental Congress, the arrival of Thomas Paine in America, the ushering in of the Age of the Machine, etc. There is no doubt that Mr. Smollin is referring to the year 1774. Nor is there any doubt that he is also referring to de Cheseaux's Comet which, with its set of six 19 million mile tails, had to be one of the finest cometary spectacles in recorded

91

TABLE 7.3   Values of the Aspect Orbs

ASPECT	ORB VALUES				
	Davison (7.15)	Hone (7.16)	Lynch (7.17)	Omarr (7.18)	Parker & Parker (7.19)
Conjunction (0°)	8°-9°	8°	10°-12°	8°-10°	8°-9°
Sextile (60°)	6°	4°	7°	8°-10°	5°-6°
Square (90°)	9°	8°	8°	8°-10°	8°-9°
Trine (120°)	9°	8°	8°	8°-10°	8°-9°
Opposition (180°)	9°	8°	10°-12°	8°-10°	8°-9°

92

TABLE 7.4   Effect Of Exact Orb On
Horoscope Shown In Figure 1.1

1. Conjunctions (0° separation)

orb = 10°		orb = 1°
Sun — Mercury	Mercury — Neptune	Jupiter — Saturn
Sun — Mars	Venus — Pluto	
Mercury — Mars	Jupiter — Saturn	

2. Sextile (60° separation)

orb = 6°		orb = 1°
Moon — Neptune	Venus — Neptune	None
Venus — Uranus		

3. Square (90° separation)

No planetary square aspects present

4. Trine (120° separation)

orb = 10°		orb = 1°
Sun — Jupiter	Mars — Jupiter	Sun — Saturn
Sun — Saturn	Mars — Saturn	
Moon — Venus	Jupiter — Neptune	
Mercury — Jupiter	Uranus — Neptune	
Mercury — Saturn	Neptune — Pluto	
Mercury — Uranus		

5. Opposition (180° separation)

orb = 10°		orb = 1°
Moon — Jupiter	Moon — Uranus	None
Moon — Saturn		

history. The one small problem with Mr. Smollin's impressive analysis is that de Cheseaux's Comet appeared in early March of 1744, and not 1774. We have no doubts, however, that Mr. Smollin could quickly come up with an equally impressive list of "historical effects" for 1744 which would once more be due to the alleged impact of de Cheseaux's Comet in that year as well.

The discoveries of the planets Uranus, Neptune, and Pluto have also provided us with an opportunity to view the empirical techniques employed in astrology. Since its discovery in 1781, Uranus has managed about two and a half 84 year sidereal revolutions about the sun, while Neptune and Pluto have yet to complete even a single lap around the sun since their discoveries in 1846 and 1930. Pluto, in fact, has barely travelled 1/6 of its orbit during its half-century stay in the store of astronomical knowledge. This limited look at the orbital motions of these planets, however, has not deterred astrologers from assigning a full complement of astrological properties to each of these remote worlds (See, for example, Davison 7.22, and Jones 7.23). The planet Pluto in fact is generally regarded as holding the astrological rulership over the sign of Scorpio, despite the fact that, since its discovery, the planet has yet to enter that particular sign.

As a final illustration of the process, let us consider the object discovered photographically by Dr. Charles T. Koval of the Hale Observatories in California. On November 1, 1977, Dr. Koval found an object 160-640 km in diameter which is moving in an eccentric orbit lying between the orbits of Saturn and Uranus. The object is most unusual in that it moves about the sun in a planet- or comet-like orbit, yet has the size of a larger asteroid. Koval has named the object Chiron, after the wisest of all the centaurs in Greek mythology who was a descendant of the gods Saturn and Uranus, and is most renown for his knowledge of medicine. In anticipation of Chiron's astrological acceptance as a full-fledged planetary influence, we have summarized some of Chiron's astrological properties in Table 7.5. Clearly, for example, Chiron's dual centaurian nature makes him a natural ruler over the sign of Gemini, and when combined with his association with medicine, Saturn, and Uranus, it is possible to generate the remaining items in Table 7.5 with minimal difficulty.

Despite such nonsense, however, there are a number of astrologers who do raise some interesting correspondences for our consideration. One of the more famous of these is the sinister "Twenty Year Sequence" of fatal misfortunes befalling United States presidents elected in years evenly divisible by 20. Excellent astrological summaries of the correspondence are given by Goodavage (7.24) and Omarr (7.25), but essentially the claim is that presidents elected at the time of Jupiter-Saturn conjunctions have died while still in office. The grim record is summarized in Table 7.6. In the discussion of the sequence, however, a number of important difficulties are almost always omitted. First of all, Thomas Jefferson was elected in 1800 at the time of a Jupiter-Saturn conjunction, was reelected in 1804, and finally died in 1826 at the age of 83, well after leaving office in 1812. James Monroe was also elected for his second term at the time of a Jupiter-Saturn conjunction in 1820 and, like Jefferson, also successfully served out that second term. Monroe then died in 1831 at the age of 73. On the other hand, Zachary Taylor who was elected in 1848 died of acute gastroenteritis on July 4, 1850, totally without the benefit of a Jupiter-Saturn

94

Figure 7.2   De Chesseaux's Comet on March 7, 1744. A comet three decades ahead of its astrological time.

Figure 7.3   Chiron (round image) in November 14, 1977. The trailed stellar images are due to Chiron's motion during the course of the exposure.

TABLE 7.5	Some Proposed Astrological Properties For Chiron
Sign Rulership	Gemini (replaces Mercury)*
Compatible Planets	Saturn, Uranus
Physical Characteristics	Powerful legs, thighs, and chest; Hair often long and mane-like; Well developed nose and nostrils.
Planetary Anatomy	Rules any body part that occurs in pairs, especially arms, legs, nostrils and lungs.
Planetary Diseases	Fractures of the limbs and shoulders; bronchitis, asthma, and emphysema.
Planetary Psychology	Strong tendency toward internal conflict caused by a fundamental split between the humanitarian and animal aspects of the personality.
Planetary Occupations	Medically related occupations, including doctors, nurses, medical technicians, and hospital workers; also jockeys, equine trainers and other occupations involving horses.
Planetary Colors	Sorrel, roan, hospital white.
Planetary Gem	Ivory
Planetary Metal	Stainless Steel

*Ends Mercury's dual rulership over Gemini and Virgo (See Chapter 11).

TABLE 7.6   The Twenty Year Sequence
For U.S. Presidents

PRESIDENT	YEAR ELECTED	FATE	DATE(S) OF JUPITER SATURN CONJUNCTIO
William Henry Harrison	1840	died of pleurisy and pnemonia July 4, 1841	January 26, 1841
Abraham Lincoln	1860	shot April 14, 1865 died April 15, 1865	October 24, 1861
James Garfield	1880	shot July 2, 1881 died Sept. 19, 1881	April 21, 1881
William McKinley	1900	shot Sept. 6, 1901 died Sept. 14, 1901	November 27, 1901
Warren Harding	1920	died of pneumonia August 2, 1923	September 14, 1921
Franklin Roosevelt	1940	died of a stroke April 12, 1945	August 15, 1940 October 11, 1940 February 20, 1941
John Kennedy	1960	shot Nov. 27, 1963 died the same day	February 18, 1961

conjunction. Moreover, one must wonder over the time dependence of the "evil" in the Jupiter-Saturn conjunction. Franklin Roosevelt's fate befell him over four years after the two planets played a three conjunction game of celestial tag in late 1940 and early 1941. Abraham Lincoln was assassinated over three years after the October, 1861 Jupiter-Saturn meeting. On the other hand, William McKinley never lived to see the November 1901 Jupiter-Saturn conjunction which supposedly catapulted Theodore Roosevelt into the Presidency.

Astrologers of course, are at the ready with explanations for such difficulties. The Jupiter-Saturn conjunctions since 1840, for example, have occurred in the so-called "earth signs",* while those prior to 1840 occurred outside of the earth sign. Therefore, the terrible effects of the Jupiter-Saturn conjunction in 1820 were nullified by the fact that the conjunction occurred in Aries, a "fire" signs. Therefore, the terrible effects of the Jupiter-Saturn conjunction in fact, the entire process of "patching up" the Twenty Year Sequence is totally analogous to the medieval process of stacking additional epicycles onto what was an already impressive array of celestial orbs in order that the insatiable idiosyncrasies of the planets' motions could be temporarily satisfied.

The Twenty Year Sequence, however, is but one of a large number of similarly interesting and seemingly impressive correspondences. In the fall of 1976, "Sports Illustrated" magazine found that in virtually every instance, for over a half century, whenever the National League won the World Series in an election year, the Democratic candidate won the Presidency; when the American League won, so did the Republicans. Using this "correspondence", the editors of "Sports Illustrated" then went on to correctly predict that Democrat Jimmy Carter would win the 1976 election on the basis of the Cincinnati Reds' four-game demolition of the New York Yankees earlier that fall (7.26).

In a similar vein, writer Erle Stanly Gardner suggested a number of years ago that the number ten was the lucky number of the infamous Willie "the Actor" Sutton, because most of his prison escapes and many of his successful bank robberies occured on the 10th day of the month. Upon hearing of this supposed correspondence, Mr. Sutton responded,

> " . . . since escapes and bank robberies depend on any number of things over which you have very little control, most particularly the weather, he would have been far better off, I could have told him, studying the magical effects of number ten on the Gulf Stream." (7.27)

In his own inimitable style, Mr. Sutton thus raises the fundamental issue of coincidence in the evaluation of a supposed correspondence. Is it coincidence or correspondence that the number ten seems to be intimately involved in Willie Sutton's finer moments? Is it coincidence or correspondence that the World Series winner is an indicator of the Presidential winner? Is it coincidence or correspondence that Jupiter-Saturn conjunctions seem to be intimately involved in our Presidential tragedies? To the astrologer, virtually *nothing* is coincidental in nature and the astrologer would therefore see some sort of significance in all

*Taurus, Virgo, and Capricornus.

of the above. To the scientist, battle-hardened and suspicious from decades of struggle with Murphy's Laws,* coincidences in nature have been found to be more the rule than the exception. Thus, in the absence of additional data, the scientist would tend to dismiss *all* of the above to coincidence. Clearly, however, some coincidences are true correspondences and behavior patterns. But which ones? Fortunately, the twentieth century has brought to us a method, modern statistics, by which we can, at least in part, answer this question. Unfortunately, the fuel needed to run the statistical method smoothly requires a numerical richness far in excess of what the bulk of the astrological correspondences can offer. Thus, when Mr. Omarr (7.28) speaks of crisis for the United States whenever Uranus crosses the sign of Gemini, the statistical method coughs to a sputtering stop when fed the three times in two centuries that such a crossing has occurred. It fares little better with the Twenty Year Sequence and its large number of statistical variables (48 elections and 39 Presidents) in comparison to the thankfully small number of in-office fatalities (8). In fact, virtually all of the early and most of the recent attempts to statistically test the astrological correspondences have suffered from problems of one sort or other. Psychoanalyst Carl Jung's oft-quoted statistical work on the horoscopes of 180 married couples, for example, was, by Jung's own admission, statistically inconclusive in light of the sample size (7.29). Addey's sample sizes (7.30) are adequate, but he then slices them up into "harmonics" which amount to little more than statistical hors d'oeuvres. One can also find, with a minimum of difficulty, presentations such as that of Helynne Hansen (7.31) whose untabulated data samples are dumped into a cavernous array of astrological variables, including planets, houses, aspects, and astrological signs. In an almost unbelievable open letter to the New York Times, astrologer A.H. Morrison presented a graph of 4,253 scientists plotted against their corresponding sun-sign, with "corroboration" from a smaller study of 102 botanists (7.32). In neither case could the distribution's "significance" withstand a proper statistical analysis. Despite such statistical mischief, however, it is possible to conduct proper statistical tests of the astrological correspondences (7.33), and, as we shall see, they will leave the empirical astrologer without any clothes.

*A whimsical set of "laws" of nature developed by experimental scientists one of which states: Nature always sides with the hidden flaw.

7.1   "Astrology: Your Daily Horoscope," March 1978, p. 99.

7.2   Agrey-Thatcher, G., "Horoscope," June, 1977, p. 46.

7.3   Culver, R., and Ianna, P., Astrophysical Journal, 195, L 37-38 (1975).

7.4   West, J. and Toonder, J., p. 19, "The Case for Astrology," Penguin Books, Inc., Baltimore, Maryland (1973).

7.5   Aahmes, N., "Astrology at Work," pp. 50-51, "Horoscope," November, 1977.

7.6   Allen, R., p. 1, "Star Names: Their Lore and Meaning," Dover Publications, Inc., New York (1963).

7.7   Ibid., Reference 7.4, p. 285.

7.8   Jobes, G. and Jobes, J., p. 44, "Outer Space: Myths, Name Meanings, Calendars," Scarecrow Press, Inc., New York (1964).

7.9   Gleadow, R., p. 163 "The Origin of the Zodiac," Atheneum Press, New York (1969).

7.10 Ibid., Reference 7.9, p. 100.

7.11 Schmidt, S., "Astrology 14," Pyramid Books, New York (1970).

7.12 Ibid., Reference 7.9, p. 18.

7.13 Ibid., Reference 7.11, p. 18.

7.14 Jenkins, F. and White, H., Chapter 19, "Fundamentals of Optics," 3rd ed., McGraw-Hill Book Company, Inc. (1957).

7.15 Davison, R., p. 110, "Astrology," ARCO Publishing Company, Inc., New York (1975).

7.16 Hone, M., p. 180, "The Modern Text-Book of Astrology," revised ed., L.N. Fowler and Company, Ltd., London (1973).

7.17 Lynch, J., p. 275, "The Coffee Table Book of Astrology," revised ed., The Viking Press, New York (1967).

7.18 Omarr, S., p. 276, "My World of Astrology," Wilshire Book Company, North Hollywood, California (1975).

7.19 Parker, D. and Parker, J., p. 132, "The Compleat Astrologer," McGraw-Hill Book Company, New York (1971).

7.20 Shulman, S., p. 21, "The Encyclopedia of Astrology," Hamlyn Publishing Group, Ltd, New York (1976).

7.21 Smollin, M., "Horoscope," November 1974, p. 21.

7.22 Ibid., Reference 7.15, pp. 69-78.

7.23 Jones, M., pp. 302-310, "Astrology: How and Why it Works," Shambhala Publications, Inc., Boulder, Colorado (1969).

7.24 Goodavage, J., p. 62, "Astrology: The Space Age Science," Signet Books, New York (1978).

7.25 Browning, N., Chapter 4, "Omarr: Astrology and the Man," Signet Books, New York (1978).

7.26 "Sports Illustrated," November 1, 1976, p. 16.

7.27 Sutton, W., (with E. Linn), p. 321, "Where the Money Was," Ballantine Books, New York (1977).

7.28 Ibid., Reference 7.18, pp. 272-273.

7.29 MacNeice, L., pp. 222-225, "Astrology," Doubleday and Company, Inc., Garden City, New York (1964).

7.30 Addey, J., "Harmonics in Astrology," Cambridge Circle, Limited, Green Bay, Wisconsin (1976).

7.31 Hansen, H., "Horoscope," February 1978, p. 59.

7.32 Morrison, A., CAO Times, 3, 4 (1977).

7.33 Dean, G. and Mather, A., "Recent Advances in Natal Astrology," The Astrological Association, Bromley Kent, England (1977).

Chapter 8
The Cosmic Vibes

Nowhere are the battle lines more sharply drawn between scientist and astrologer than over the issue of astrological planetary influences. Even the astrologers themselves are in sharp disagreement over the workings of these alleged influences. Historically the astrological view has been that these influences are causal in nature, i.e. the planet Venus *does* something that causes or brings about the observed astrological effects. Over the past three centuries, scientists have discovered a variety of celestial agents such as gravity, electromagnetic radiation, etc. which can affect the earth in one way or other. The properties and behavior of all of these known celestial influences have been painstakingly deduced and established from the results of thousands of observations and experiments made by scientists during the past several centuries.

A similar but centuries-older description, based on the results of thousands of horoscopes cast down through the ages, has been developed by astrologers for the astrological planetary influences. The problem that instantly appears, however, is that the properties set forth by the astrologers for their influences are almost totally inconsistent with those discovered by modern empirical science.

Gravity, for example, is a universe-pervading influence in which every mass in the universe attracts every other mass with a force which behaves as follows:

$$\text{Force of Gravity} = G \left\{ \frac{\text{Mass 1} \times \text{Mass 2}}{(\text{Distance})^2} \right\}$$

where the quantity G is the universal gravitational constant. This law has been developed on a strictly empirical basis and has provided scientists with a multitude of experimentally verifiable predictions regarding the behavior of masses in motion in the universe (one of which is the prediction that objects of unequal mass will fall to the ground at exactly the same rate). It also tells us that the force or influence due to gravity depends *only* on three items: the individual masses of the two interacting objects, and distance between them. How then does this empirical result compare with the properties of the astrologers' cosmic vibrations?

This question can best be answered by calculating the various planetary gravitational forces which will be acting on a child at the time of birth. In Table 8.1 all of the planetary gravitational forces are computed relative to the force of gravity between the child and the planet Mars.* In addition the gravitational forces for a few objects not deemed astrologically significant are also included in Table 8.1. The resulting array of gravitational forces provide us with some interesting insights into the situation.

First of all, the planet Pluto is supposed to exert significant planetary influences on the charts of individuals, yet we see for example that the satellites Titan of Saturn, and Ganymede and Callisto of Jupiter exert respectively about 20, 80, and 50 times as much gravitational force as does Pluto, yet nowhere are any of these satellites even mentioned in the casting and interpreting of horoscopes. Even more interesting is the fact that the force of gravity between the mother and child at the time of birth is exceeded only by the gravitational forces of the sun, moon, Venus, and Jupiter, and the force of gravity between the center of mass of the hospital building itself and the child is exceeded only by the gravitational forces of the sun and the moon. Sets of reasonable assumptions other than those listed in Table 8.1 are found to yield basically the same results.

Thus if the astrologer is to cast a horoscope in which gravity is assumed to be the influencing agent, then not only must he or she take into account the array at birth of many of the lesser bodies in the solar system, but of the array of masses in the immediate vicinity of the child as well. Failure of astrologers to cast horoscopes in such a fashion can therefore only be interpreted as a repudiation of gravity as the acting agent.

Empirical science's tidal forces seem to hold a particularly strong fascination for the astrologers. Linda Goodman (8.1) typifies the attempt to leap the chasm with tides:

"Science recognizes the Moon's power to move great bodies of water. Since man himself consists of seventy percent (sic) water, why should he be immune to such forceful planetary pulls?"

But tidal forces exerted on a given object (Mass 1) by another (Mass 2) are known to arise from the difference in the force of gravity between the near and far sides of the given object, and as such, have a well-studied behavior which can be described as follows:

*The choice of Mars here as the standard is strictly arbitrary. Because the planet is deemed astrologically significant, it does provide us however with a useful point of reference.

TABLE 8.1   Gravitational Forces Acting On A Child At Birth

GRAVITATIONAL FORCE* FROM:	VALUE RELATIVE TO MARS' FORCE OF GRAVITY
Mother	20
Doctor	6
Hospital Building	500
Sun	854,000
Moon	4600
Mercury	.38
Venus	27
Mars	1
Ceres	.00008
Jupiter	46
Ganymede	.004
Callisto	.002
Saturn	3.3
Titan	.008
Uranus	0.1
Neptune	.03
Pluto	3.1×10^{-6}
A Comet	$<.00001$

*Assumptions

All celestial objects are at closest approach to Earth
Mass of Child = 3 kg.
Mass of Mother = 50 kg.
Mass of Doctor = 75 kg.
Mass of Hospital Building = 2.1×10^6 kg.
Distance between Mother and Child = 0.15 m.
Distance between Doctor and Child = 0.3 m.
Distance between Hospital Building center of mass and Child = 6.1 m.

105

Figure 8.1

A comparison of "comic" versus terrestrial influences due to gravity. The mother, doctor, and hospital building exert respectively 20, 6 and 500 times as much gravitational force on a child at birth than does the astrologically important planet Mars.

$$\begin{matrix} \text{Tidal force} \\ \text{of Mass 2 on} \\ \text{Mass 1} \end{matrix} = \left\{ \frac{2G\,(\text{Mass 2} \times \text{Size of Body 1})}{(\text{Distance})^3} \right\}$$

If we calculate the tidal forces on the child at birth using our previous set of assumptions, we obtain the results listed in Table 8.2, none of which are particularly gratifying for detente on this issue. In looking at the tidal forces exerted by the various entities in Table 8.2, we now find that the tidal forces exerted by the people and objects in the immediate vicinity of the child totally overwhelm those exerted by all celestial objects, even those old and storied tide raisers, the sun and moon. Moreover, astrologically insignificant Ceres, as well as a few of its larger colleagues in the asteroid belt have now crept up to and passed Pluto on the tidal influence scale. Thus the assumption that tidal forces are responsible for the astrological planetary influences requires an even greater emphasis on the array of nearby terrestrial mass points and lesser planetary objects in a "tidal force" horoscope than would be required for a "gravitational force" horoscope. Since we have already seen that such object arrays are *not* considered in the erection of a chart, tidal forces must therefore join gravity in the astrological reject pile.

Yet another of the scientifically describable celestial influences is the electromagnetic energy emanating from the sun, moon, and planets. Such radiant energy can be viewed as a likely candidate for the supposed astrological planetary influences. The Parkers (8.2), for example, state:

"Why does astrology work? ... There seem to be no purely logical reasons to suppose that the (planet's) positions in themselves can provide a full answer to the problem. It is much more likely that we must turn to radiations of some sort of other".

In their investigations of these radiations, scientists have, from the results of thousands of experiments, firmly established the fact that the light emitted from both celestial and terrestrial sources behaves in exactly the same fashion. And so if the properties are the same for terrestrial electromagnetic radiation as they are for the celestial electromagnetic radiation, what impact would this have on an "electromagnetic" horoscope.

If we were to place a single 200 watt light bulb in the delivery room 2 meters away from a child at the time of birth, the intensity of the electromagnetic energy impinging on the child due to that bulb is exceeded only by that coming from the sun. Moreover, the brightness of some of the fainter, astrologically significant objects is exceeded by the brightnesses of objects that are deemed to be astrologically unimportant. For example, 22 planetary satellites in the solar system are brighter than the planet Pluto and two more have the same brightness. Comets enter the solar system at an average rate of about five or so per year, and virtually all of them are brighter than Pluto at their perihelion points. From the data in "Tables of Minor Planets" (8.3), 1209 asteroids have brightnesses as bright or brighter than does Pluto. The zodiacal light and gegenschein are radiant phenomena which arise from interplanetary space and

TABLE 8.2   Tidal Forces Acting On A Child At Birth

TIDAL FORCE* FROM:	VALUE RELATIVE TO MARS' TIDAL FORCE
Mother	1.1×10^{13}
Doctor	2.0×10^{12}
Hospital Building	7.0×10^{12}
Sun	448,000
Moon	931,000
Mercury	0.3
Venus	52
Mars	1
Ceres	0.00005
Jupiter	5.8
Ganymede	0.0005
Callisto	0.0003
Saturn	0.2
Titan	0.00005
Uranus	0.003
Neptune	.0004
Pluto	4×10^{-8}
A Comet	$<10^{-9}$

*Assumptions are the same as for Table 8.1

which are not only visible to the naked eye, but also appear to creep along the zodiac just as surely as the sun, moon, and planets.

Nowhere could we find a horoscope which considered the impact of configuration of the delivery room lights, positions of faint asteroids, or faint periodic comets near perihelion. Exit the known electromagnetic spectrum as a possible source of the astrological influences.

At this point, astrologers often suggest that the influences may be due to an as yet undiscovered form of electromagnetic energy, pointing out that only a century ago, science had no knowledge of gamma rays, x-rays, radio waves, etc. In this vein the Parkers (8.4) continue:

"The essential point behind this chain of argument is that we cannot pretend to know about the various emissions pervading the universe. Some new kinds have been discovered in recent years . . . Before these revelations it was only too easy to dismiss astrology as being false simply because there were no obvious emissions or vibrations which could account for it. Today such a claim would be most unwise."

The beauty of the concept of the electromagnetic energy spectrum is, of course, that all of its forms of radiant energy can be described by exactly the same mathematical model that is used to describe visible light. Not a single one of the newly discovered forms of radiant energy mentioned by the Parkers have failed to satisfy these descriptions, and at present there is no evidence to support the idea that undiscovered varieties of electromagnetic energy lying outside of the currently known wavelength limits will have fundamentally different properties than their brethren which lie within.

If we move on to the realm of magnetism, we find that some of the planets do indeed exhibit magnetic fields. The earth, for example, has a general magnetic field which is capable of producing various magnetically-related effects such as compass needle alignments and the Van Allen radiation belts. The sun has long been known to harbor very complicated magnetic activity which makes itself manifest in sunspots, solar flares, prominences, and a wide variety of other solar phenomena. Observations made from the Mariner and Pioneer spacecraft have established the existence of magnetic fields about the planets Mercury and Jupiter as well. Unlike gravitational, tidal, and electromagnetic effects, all of which are exerted to some degree by each and every object in the solar system, there are some planets which have not been blessed with magnetic fields. Space missions to the moon, Venus and Mars, for example, reveal that these objects have no detectable magnetic fields. The list might be well expanded in future years as we extend our knowledge of planetary magnetic fields to the planets Saturn, Uranus, Neptune, and Pluto. On the basis of current data, however, we are able to say that any horoscope based on magnetic forces should have the moon, Venus, and Mars deleted as contributing factors. We have been unable to turn up an astrologer willing to cast such a horoscope.

The situation deteriorates even further if we consider the particle emissions which impinge on the earth from space. The problem between scientist and astrologer here is the fact that the only particle emitter of consequence in the entire solar system is the sun itself. Not one shred of observational evidence

TABLE 8.3   Electromagnetic Energies Incident On A Child At Birth

OBJECT	ELECTROMAGNETIC ENERGY RELATIVE TO THAT COMING FROM MARS
Sun	3×10^9
200 watt bulb*	9×10^6
Full moon	7,600
Mercury	0.4
Venus	4.4
Mars	1
Jupiter	0.8
Saturn	0.1
Uranus	.0004
Neptune	.00005
Pluto	8×10^{-8}

*Assumed distance from child = 2 meters.

gleaned from either earth-based or space probe instrumentation even hints otherwise. Having the sun standing as the sole particle emitter in the solar system may be of comfort to the pure sun-sign astrologers (see Chapter 9), but astrological horoscopes, without exception, include not only the sun, but the moon and all of the other planets as well. Other forces such as electrostatic and nuclear forces are known to exist in nature, but any horoscope cast within the constraints of these forces' observed properties would once more take on characteristics completely alien to the current methodology of astrological chart erection.

Thus the properties of the astrological planetary influences, are not compatible with those of any forces or influences currently known to the scientist. Instead, suppose, as many astrologers suggest, that we are dealing with some sort of an "influence" phenomenon that is not yet known to science. From a scientific viewpoint, then it would be extremely useful to assemble a list of the behavioral properties of such a force based solely on the observations of the astrologers themselves. Unfortunately, nowhere does the pluralistic nature of the astrological world make itself more manifest than over this particular question. Since we do not enjoy an unlimited amount of space with which to deal with all of the astrological viewpoints, we will instead attempt to summarize those properties of the planetary influences about which there is a reasonable amount of agreement among astrologers.

In general, no attempt is made among astrologers to include a planet's size, mass, density, rotation rate or any other physical property in a horoscope. We can only conclude from this aspect of the astrologers' methodology that the physical characteristics of a given planet are not regarded by astrologers as being astrologically important. This viewpoint is further strengthened by the fact that there have been very few, if any changes in the astrological properties of the planets during the very time when our space missions have brought about some rather striking changes in much of our knowledge of *astronomical* planetary characteristics. For example, the astrological response to the discovery that the planet Mercury possesses a magnetic field and spins on its axis once every 59 days instead of 88 days was essentially nil. Despite this lack of correspondence between the astrological planetary influences and the physical properties of the planets, each planet is said to exert its own peculiar astrological influence which is different from that of any of the other planets in the horoscope.

Possible relationships between the properties of a planet's orbital motion and its astrological influences are ill-defined at best. As we shall see, astrologers generally claim that the three outermost planets, Uranus, Neptune, and Pluto affect entire generations because of their relatively slow angular motion through the astrological signs. Retrograde motions, the apparent "backward" east to west motions of the planets, are generally regarded as being both astrologically significant and astrologically different from the normal direct or west to east apparent motions of the planets. The exact nature of such relationships between the astrological planetary influences and a planet's apparent motion however seems to be a matter of considerable disagreement among astrologers.

The spatial behavior attributed by astrologers to the planetary influences is particularly intriguing. In order to consider these qualities from an astrological viewpoint, it is useful to employ a coordinate system that uses the ecliptic as its fundamental reference circle. A three-dimensional description of an object's position in space relative to that of the earth is then accomplished by the use of three coordinates, the "celestial latitude" or angular distance north or south of the ecliptic, the "celestial longitude" or eastward angular distance along the ecliptic from the vernal equinox point, and the line of sight between the earth and the object.

We have already seen that the astrologer attaches a great deal of significance to the angular configurations or aspects of the planets in evaluating the impact and importance of planetary influences. In virtually every astrology text consulted regarding the casting of horoscopes, however, the planetary aspects are measured as angular separations *along the ecliptic,* i.e., angular separations in celestial longitude. For example, on September 6, 1975, the moon, Pluto, and Mercury were lined up in Virgo as shown in Figure 8.2. Because the celestial longitudes of these objects were within the allowed orbs of each other, astrologers interpreted this arrangement as a triple conjunction of the moon, Pluto, and Mercury. Actually, the angular distances in two dimensions were such that Pluto was 18° from Mercury and 21° from the moon and hence was outside of the allowed orb in celestial latitude for a conjunctive aspect to occur. Thus, Pluto should not have aspected either Mercury or the moon if celestial latitude were a factor in computing planetary aspects. It should be noted that brief homage is paid in a few texts to aspect separations in celestial latitude and these latitude aspects, called "parallels", are used by some astrologers, to interpret their clients' horoscopes. Judging from the techniques currently in print for casting the horoscopes as well as the tables of planetary positions (which usually omit celestial latitude data), however, we must conclude that parallel aspects are usually excluded from the mainstream of astrological horoscope analyses.

The line of sight distance, like the celestial latitude, does not seem to enter into the interpretation of a horoscope in any consistent fashion. For example, the astrologically meaningful properties of a given aspect involve the type of aspect (conjunction, opposition, etc.), the planets involved, and whether or not the aspect is "applying" (angular separation decreasing in time) or "separating" (angular separation increasing in time). Of particular interest here are the conjunctions between Mercury and the sun and Venus and the sun. Because both of these planets move about the sun in orbital paths that lie within the earth's orbit, it is possible for these planets to exhibit two basic types of conjunction with the sun, the so-called "inferior" conjunction when the planet lies between the earth and the sun and "superior" conjunction when the planet lies on the far side of the sun as seen from earth (Figure 8.3). To the best of our knowledge very few astrologers make the distinction between inferior and superior conjunctions. Certainly West and Toonder don't (8.5), nor do Sydney Omarr (8.6) or Margaret Hone (8.7). Even the venerated Grant Lewi (8.8) seems to ignore the issue. If we are considering line of sight distances, however, the distinction is a most important one, for Mercury is about 2.3 times closer to the earth at inferior conjunction than at its superior conjunction, and Venus is 6.1

112

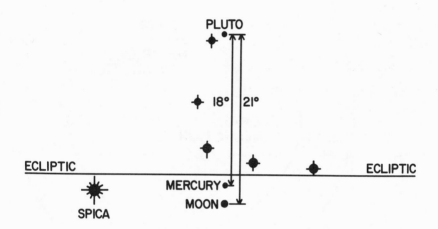

Figure 8.2

The configuration of the Moon, Mercury, and Pluto on September 6, 1975 in the constellation of Virgo. Although this is generally regarded by astrologers as a triple conjunction of planets, Pluto's vertical angular distance lies outside of the orbs of Mercury and Moon.

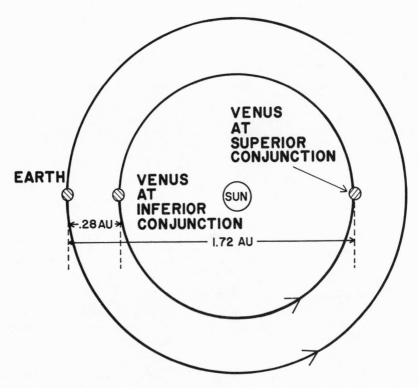

Figure 8.3

Inferior and superior conjunctions of the planet Venus. At inferior conjunction Venus is more than six times closer to the earth than it is at superior conjunction. Distances are expressed in terms of the mean radius of the earth's orbit or "astronomical unit" (AU).

times closer to the earth at its inferior conjunction than at its superior conjunction.

The issue of the effect of line of sight distance, thus, seems to be one that is at best treated in a very superficial and qualitative fashion by astrologers, and in the absence of a quantitative mathematical expression relating the observed astrological intensity of planetary influences to line of sight distance, we will assume that none exists.

For the most part, then, astrologers evaluate the planetary configurations in a horoscope in terms of differences in celestial longitude only. The exact meaning or impact of each aspect depends astrologically on the planets involved and the astrological signs in which they occur. Astrologers generally consider the conjunction, opposition, square, trine, and sextile as powerful and significant aspects. The square and opposition are viewed as "bad", "unfortunate", or "disturbing" aspects while the sextile and trine are regarded as "good", "fortunate", or "harmonious" aspects. Conjunctions can be harmonious or disturbing, depending on the planets involved. The orb associated with a given aspect essentially represents the zone over which that aspect's influence will be felt. If we assume an opposition orb of $\pm10°$, opposition influences will thus be active over separations in celestial longitude ranging from $170°$ to $190°$. There is a considerable amount of disagreement in the astrology literature over the behavior of the planetary influences over the orb interval (8.9). One of the more bizarre ideas is the concept of the combust which according to Hall (8.10)

"When planets are within $3°$ of the sun, their influence is partly destroyed by the solar forces. This position is very unfortunate."

The most common, "first-order" approximation however is that the planetary influences behave more or less like step functions: full strength influence inside of the aspect orb limits, and zero elsewhere. With this information, we can then construct the qualitative behavior pattern shown in Figure 8.4 in which the influence of an aspect is plotted as a function of the angular separation of the two planets involved. The inclusion of the "lesser" aspects such as $45°$ and $135°$ has the effect of putting more "teeth" into the diagram, but does not change the basics of the results. The orbs assumed for Figure 8.4 are from Parker and Parker (8.11). The use of a different set of orbs either fattens up or slims down the aspect zones, but again does not alter the overall nature of the plot shown in Figure 8.4. We now have at least a qualitative picture of how the astrological planetary influences behave as a function of angular separation.

The property of the astrological planetary influences perhaps most at odds with current scientific thought is the coupling effect that astrologers contend exists between the planets and the astrological signs in which they are located. In the ancient and medieval geocentric models for the planetary system, the final, outermost sphere, the "star sphere" was given over to the fixed stars. Since the stars' distances were assumed to be of the same order of magnitude as the planetary system, the idea that the star sphere could interact with the planetary spheres was a perfectly reasonable one. As we have seen, the background signs onto which the sun, moon and planets are projected in a horoscope are most

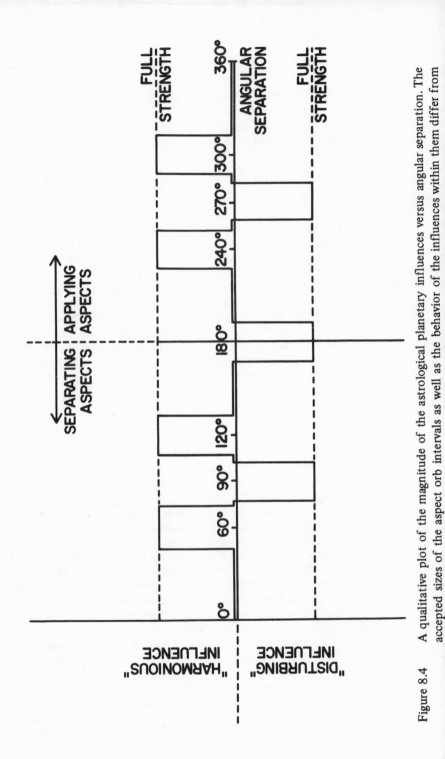

Figure 8.4   A qualitative plot of the magnitude of the astrological planetary influences versus angular separation. The accepted sizes of the aspect orb intervals as well as the behavior of the influences within them differ from

significant from an astrological viewpoint. Especially important are the exact locations on the ecliptic of the ascendant, descendant midheaven, astrological nadir, and their midpoints, since each of these points can be aspected by the sun, moon, and planets. Aspects to the midheaven, for example:

"... will affect the self-expression in the outer life, hence influencing prestige, standing, and the career; also the relationship to those in authority and sometimes to a parent, more probably the father ... (8.12).

The ascendant, midheaven, descendant and astrological nadir and their midpoints are thus thought of as being imprinted onto the zodiac at the time of the individual's birth and remain astrologically significant in any progressed charts cast later in that person's life. In a similar vein, many astrologers employ one or more of the so-called "astrological parts" which are imaginary points in the horoscope that are defined in terms of known celestial longitudes in the horoscope (see Table 8.4). None of these points, however, bears any direct relationship to any celestial object, nor do they possess any *astronomical* significance as do the ecliptic's equinox and solstice points or the nodal points of the lunar orbit.

Some astrologers make use of the fixed stars in the interpretation of a chart. Hone, for example, tells us:

"In a natal chart, if any planet is on the degree of an important fixed star, it seems to be strengthened and to act more strongly in combination with the nature of that star." (8.13)

And astrologer Manly Hall echos the concept:

"Fixed stars are most powerful in conjunction and parallel, but are felt in square and opposition ... their orb of influence is 5°. Only those fixed stars should be considered which conjunct planets or are on the cusps of houses." (8.14)

Few contemporary astrologers, however, seem eager to embrace this viewpoint, since it clearly produces a fearsome array of philosophical liabilities in view of what we know about the vastness of stellar distances relative to planetary distances. Nonetheless, emerging from all of this is the basic astrological concept that zodiacal space is imprintable and possesses influences of its own which are capable of astrological interactions with the sun, moon, and planets.

In summary, then, from the astrological literature itself, we can assign the following properties to the astrological planetary influences:
(1) The influences do not arise from any force or influence presently known to science, including gravitational and tidal forces, electromagnetic radiation, magnetic fields, or particle emissions.
(2) The influences bear no correlation to any intrinsic physical property of a given planet, including mass, size, density, rotation rate, surface and atmospheric composition, yet they are uniquely different for the Sun, Moon and each of the planets.

117

TABLE 8.4   A Partial Table Of The Astrological Parts (8.15)

ASTROLOGICAL PART	LOCATION: Celestial longitudes of:
Fortune	Ascendant + Moon − Sun
Commerce	Ascendant + Mercury − Sun
Love	Ascendant + Venus − Sun
Passion	Ascendant + Mars − Sun
Increase	Ascendant + Jupiter − Sun
Fatality	Ascendant + Saturn − Sun
Catastrophe	Ascendant + Uranus − Sun
Treachery	Ascendant + Neptune − Sun
Organization	Ascendant + Pluto − Sun
Spirit	Ascendant + Sun − Moon
Faith	Ascendant + Mercury − Moon
Female Children	Ascendant + Venus − Moon
Male Children	Ascendant + Jupiter − Moon
Merchandise	Ascendant + Fortune − Spirit
Servants	Ascendant + Moon − Mercury
Understanding,	Ascendant + Mars − Mercury
Mother	Ascendant + Moon − Venus
Father	Ascendant + Sun − Saturn
Brothers & Sisters	Ascendant + Jupiter − Saturn
Water Journeys	Ascendant + Cancer 15° − Saturn
Play	Ascendant + Venus − Mars
Discord	Ascendant + Jupiter − Mars
Inheritance	Ascendant + Moon − Saturn
Sickness	Ascendant + Mars − Saturn
Sudden Advancement	Ascendant + Sun − Saturn
Life	Ascendant + Moon − Full Moon Next Before Birth

(3) Aside from a separation of the non-generation and generation planets (see chapter 11) and the astrological qualities assigned to planets in direct and retrograde motion, the various aspects of the orbital motions of the planets do not appear to enter into the astrologers' evaluations of the planetary influences. Such motions are, of course, necessary in casting a horoscope and the progressions of the horoscope.

(4) The influences are independent of the line of sight planetary distances.

(5) The influences are independent of the planetary angular distances north or south of the ecliptic plane.

(6) The influences are related to the relative positions of the planets along the ecliptic. Important angular separations include 0°, 60°, 90°, 120°, and 180°. Within the orbs of these angles, the influence of the aspect is present and is distributed over the orb interval in some not universally agreed on fashion; outside of the orb the influence is zero.

(7) Astrological space possesses a set of astrological influences of its own. Moreover, this space can be imprinted with crucial astrological points, such as the midheaven and the ascendant. The distant stars are generally regarded as having no effect on the zodiacal space or its influences.

Very quickly we can see that the description of the behavior of the astrological planetary influences as set forth by the astrologers themselves generates a set of constraints that the scientists find overwhelmingly incompatible with their empirically determined descriptions of the universe.

The idea that space can be imprinted and exert influences of its own, for example, leads us to some interesting questions regarding the possible impact on an imprintable zodiacal space arising from the millions of imprinted ascendant points, midheavens, etc. that must have occurred since the beginnings of humanity on the planet, let alone the animals, nations, organizations, etc., for which horoscopes are deemed significant. As we noted above, the idea of an imprintable space is quite consistent with the ancient and medieval concepts of crystalline spheres, but the crystalline spheres were shattered by Tycho's observations of the Great Comet of 1577 (8.16), and their imprintable replacement, the interstellar ether, was evaporated by the results of the Michelson-Morley experiment in 1881 (8.17, 8.18). The literal vacuum left behind in the wake of these observational results is not, from a scientific viewpoint, an imprintable medium.

At this point the science-astrology discussion often takes on a pre-occupation with the "crucial argument" that would gladden the heart of any medieval scientist. On one hand the scientist argues that "astrology cannot work, therefore it does not work". The astrologers respond in one of several ways. One of the more common claims is that since "astrology works", there *must* exist a planetary influence which is as yet undiscovered by science and which possesses the aforementioned astrological properties. This is not so far fetched, goes the argument, if we remember that there are perfectly accepted scientific concepts in relativity theory and quantum theory which are also seemingly contrary to our everyday experience. What is often forgotten, however, is that both quantum theory and relativity theory are also scientifically accountable for the observed results of the "everyday" world. In other words, these theories must, in a single description, bridge the gap between the marvelous phenomena of relativistic and quantum physics and the more familiar and mundane phenomena of the "everyday" physics. Neither astrologer nor scientist has been able to perform this feat with the astrological planetary influences and their nightmarish set of behavioral properties enumerated above. As a result, many contemporary astrologers have abandoned the idea that the sun, moon, and planets astrologically cause or in some way precipitate human events and interactions, and have instead embraced "non-causal" explanations for the workings of astrology.

Currently, the most popular of the non-causal explanations for the astrological influences is the concept of "synchronicity". Basically synchronicity makes the claim that there is a correspondence or a coincidence between various astrological planetary configurations and various aspects of human activity here on earth, or in the familiar astrological refrain, "As above, so below". Suppose, for example, we observe that the rising of a certain bright star or prominent constellation exactly at sunset coincides with the first day of spring. Now the rising of that star or constellation does not *cause* the start of springtime, but is rather an occurence in the heavens which coincides with that terrestrial event. The rising can hence be interpreted as a celestial sign which corresponds to or is synchronous with the start of spring. Although astrologers usually attribute the idea of astrological synchronicity to the Swiss psychiatrist Carl Jung, the basic idea is, as we have seen earlier, virtually as old as humanity itself, as witnessed by Stonehenge (8.19), the Cheyenne Medicine Wheel (8.20), and numerous other human artifacts and monuments throughout the world.

Ideally any such astrological correspondences should be of the "lead pipe cinch" variety in which each and every time the celestial signs occur, the corresponding terrestrial event occurs. Obviously, however, in dealing with human beings, such strict one-to-one correspondences are simply non-existent. The best next requirement for a useful astrological correspondence then is one which occurs a significant fraction of the time, or in other words, a *statistically significant* fraction of the time. With the astrological retreat from the idea of causal planetary influences, be they from ancient gods and goddesses, or from more modern forces such as gravity, investigation of the cosmic vibes has been synchronously plunged into the often murky work of statistical interpretation, whose methods and techniques offer the investigator a marvelous spectrum of opportunity ranging from a chance to solve empirical problems that are otherwise insoluble to, alas, a nearly limitless capacity for interpretive mischief and disaster.

8.1   Goodman, L., "Linda Goodman's Sun-Signs", p. 477, Bantam Books, Inc., New York (1968).

8.2   Parker, D. and Parker, J., "The Compleat Astrologer", p. 52, McGraw-Hill Book Company, New York (1971).

8.3   Pilcher, F. and Meeus, J., "Tables of Minor Planets", Illinois College, Jacksonville, Illinois (1973).

8.4   Ibid., Ref. 8.2, p. 52.

8.5   West, J. and Toonder, J., "The Case for Astrology", p. 127, Penguin Books, Inc., Baltimore, Maryland (1973).

8.6   Omarr, S., "My World of Astrology", Wilshire Book Company, North Hollywood, California (1975).

8.7   Hone, M., "The Modern Text-Book of Astrology", p. 186, revised ed., L.N. Fowler and Company, Ltd., London (1973).

8.8   Lewi, G., "Astrology for the Millions", p. 124, 4th ed., revised, Llewellyn Publications, St. Paul, Minnesota, 55165, U.S.A.

8.9   Dean, G., and Mather, A., "Recent Advances in Natal Astrology", pp. 337-339, The Astrological Association, Bromley Kent, England (1977).

8.10   Hall, M., "Astrological Keywords", p. 94, Littlefield, Adams and Co., Totowa, New Jersey (1958).

8.11   Ibid., Ref. 8.2, p. 132.

8.12   Ibid., Ref. 8.7, p. 199.

8.13   Ibid., Ref. 8.7, p. 280.

8.14   Ibid., Ref. 8.9, p. 175.

8.15   DeVore, N., "Encyclopedia of Astrology", pp. 14-15, Littlefield, Adams & Co., Totowa, New Jersey (1976).

8.16   Gingerich, O., *Sky & Telescope*, **54**, 452 (1977).

8.17   Abers, E. and Kennel, C., "Mater in Motion", pp. 357-363, Allyn and Bacon, Inc., Boston (1977).

8.18   Shankland, R.S., *Scientific American*, Nov., 1964, p. 107.

8.19   Hawkins, G., "Stonehenge Decoded", Doubleday Company, New York (1965).

8.20   Eddy, J., *Science*, **184**, 1035 (1974).

Chapter 9
The Great Sun-Sign Myth

No concept in astrology is more universal and pervading than the idea that human traits are, to a large extent, determined by an individual's sun-sign. Even the most casual glance at the astrological literature reveals a constant classification of individuals strictly according to their respective sun-signs. In her best-selling astrology book, "Linda Goodman's Sun-Signs", Linda Goodman perhaps best summarizes the supposed sun-sign effect:

> "The sun is the most powerful of all the celestial bodies. It colors the personality so strongly that an amazingly accurate picture can be given of the individual who was born when it was exercising its power through the known and predictable influence of a certain astrological sign". (9.1)

Various astrological authors disagree on just how much of an "amazingly accurate picture" can be deduced from a person's sun-sign, but most are agreed that correlations do exist between a person's sun-sign and various aspects of that person's existence. Depending on the astrological source consulted, the sun-sign is supposedly related to one or more of the following components of one's make-up: personality traits, physical appearance, physiological characteristics, career, health problems, and even compatibility with other persons.

The concepts of sun-sign astrology are thus very simple, which no doubt contributes heavily to their popularity with the general public. At the same time, however, it is this very simplicity that permits us to also empirically test them with relative ease. In principle, all we have to do is sort out a sample of people according to their sun-sign and see if the alleged correlations hold up. To illustrate the point, one of the supposed sun-sign influences is involved with individual physical appearance. Hone (9.2), for example, tells us that the physical characteristics of an Aries person include:

122

"A longish stringy neck like a sheep; A look of the symbol   in the formation of eyebrows and nose; well-marked eyebrows; Ruddy complexion; Red hair, active walk."

Clearly not all people born under the sign of Aries have red hair and "ruddy" complexions. This fact is easily demonstrable by simply considering a sample of people who not only are born under the sign of Aries, but possess a Black, Oriental, or Hispanic ethnic background as well.

On the other hand, we can ask whether a sample consisting only of red-haired individuals will exhibit Aries as the predominate sun-sign. To test this possibility a survey was conducted on 300 red-haired people in which each person was categorized according to that person's sun-sign. The results of that survey are presented in Table 9.1. From these data, two things are immediately obvious. First of all, the sun-sign Aries has no special significance in terms of the number of red-headed people born under this sign. Secondly, *no* sun sign enjoys the striking, stand-out type of superiority that could immediately be translated into a scientific law. For example, if 250 of the 300 people in the survey had been Arians, we would strongly suspect the existence of an Aries-red hair correlation.

If we look at the data in Table 9.1, however, we do find that some of the sun-sign bins contain more people than others. The sign of Libra, for example, has the largest number of red-haired individuals with 34. At this point it is tempting to claim that the sign of Libra is correlated with red-haired people. But do these 34 Librans exhibit any statistical significance when compared with the sample sizes born under the remaining eleven sun-signs? In particular, what are the chances that we can get 34 Librans with red hair from a totally random sun-sign distribution of 300 red-haired people? We know that one hundred tosses of an honest coin most often do not produce exactly 50 heads and 50 tails, but rather a heads-tails split that is somewhat different, such as 52 heads and 48 tails, etc. In like fashion 60 rolls of a single unloaded die seldom result in exactly 10 appearances of each side of the die. To what extent, then, can totally random processes produce distributions which are not precisely and uniformly random? The statisticians have developed a number of sophisticated techniques by which this question can be answered.

One of the most powerful and commonly used of these tests is a procedure known as the chi-square test.* Briefly, the X^2-test allows us to compare an observed distribution of data with some sort of an assumed distribution by calculating the probability or "odds" that one would obtain the observed distribution of data instead of the assumed or "standard" distribution. There is almost universal agreement on the part of statisticians that the observed and assumed distributions of a data set are statistically indistinguishable if the calculated X^2 probability is larger than .05, or 20 to 1 odds. In the minds of many scientists and statisticians, however, Nature is sufficiently wiley and cantankerous to warrant the use of tougher criteria, and so quite often departures from an assumed distribution are not deemed statistically significant

*Usually written as X^2-test.

123

TABLE 9.1 Observed Sun-Signs of 300 Red-Haired People

SUN-SIGN	NUMBER OF PEOPLE	SUN-SIGN	NUMBER OF PEOPLE
Aries	27	Libra	34
Taurus	22	Scorpio	23
Gemini	25	Sagittarius	21
Cancer	28	Capricorn	31
Leo	30	Aquarius	20
Virgo	21	Pisces	18

unless the odds are at least 1000 to 1 against obtaining the observed data set instead of the assumed set.

To illustrate the process, let us return to Table 9.1 and test that data set to see if it is statistically different from a totally random sun-sign distribution of red-haired people. If the sun-sign distribution is totally random, we would expect 300 ÷ 12 or 25 red-haired people to be born under each sun-sign.* The X^2 probability obtained** from the comparison of the Table 9.1 distribution of red-haired people by sun-sign is statistically consistent with the set of observed data listed in Table 9.1. Hence there is no reason to believe that a statistically significant correlation exists between red-haired people and *any* of the sun-signs, including Aries.

On the other hand, if 250 of the 300 people in our survey had been born under the sun-sign of Aries, then the X^2 probability would be about 10^{-18}, a convincing repudiation of the assumption of a random distribution by anyone's standards! Similar statistical tests have been applied at various times by different investigators to some of the claims made by sun-sign astrology, and it is most interesting to consider not only their results but also the astrological response to them.

Historically a large number of astrologers have claimed that the physical characteristics of a given person are determined to a large degree by that person's sun-sign. Hone is not alone in her claims for a relationship between sun-sign and physical characteristics, and while this concept has been abandoned to a certain extent, the current astrological literature still abounds with descriptions of the physical characteristics of the "pure" sun-sign types. Shulman (9.5) for example describes the Arian characteristics as follows:

"a dry body, not exceeding in height; lean or spare, but lusty bones, and his limbs strong; the visage long, black eye-brows, a long scraggy neck, thick shoulders; the complexion dusky, brown or swarthy."

And Higgins (9.6) portrays Arians as having

" . . . a quick darting appearance, rather a lean and wiry torso, and a long, slender neck. Its subjects are often round headed and snub-nosed, with a prominent chin. The teeth are quite prominent, the eyes usually hard or grey, and the hair is reddish or light brown, usually curly or wiry."

In fact, the statistical evidence currently available on the matter emphatically demonstrates that there is no basis to the astrological claim that an individual's physical characteristics are correlated to that person's sun-sign in any way. Based on a survey conducted at Colorado State University, some of those physical

---

*Actually, because the sun does not spend exactly the same number of days in each sign during the course of the year, the assumed distribution will not be constant with sun-sign. More difficult is the problem of seasonal variation in births which has been discussed more fully elsewhere (9.3, 9.4).

**The details of the X^2-test will not be recounted here, but can be obtained from any basic statistics book.

TABLE 9.2   Physical Characteristics Not Correlated to Sun-Sign

Arm size (women)	Height (women)
Baldness	Hip size (men)
Bicep size (men)	Hip size (women)
Blood type	Lefthandedness
Bust size (women)	Leg length (men)
Calf size (men)	Leg length (women)
Chest size (men)	Neck size (men)
Finger size* (men)	Reach (men)
Finger size* (women)	Righthandedness
Fist size (men)	Sex
Foot size (men)	Skin color
Foot size (women)	Waist (men)
Freckles	Waist (women)
Hair color	Wrist size (men)
Hand size (women)	Wrist size (women)
Head size (men)	Weight (men)
Head size (women)	Weight (women)
Height (men)	

*ring finger

characteristics found not to be correlated to sun-sign are listed in Table 9.2. Each category in Table 9.2 had a sample size of at least 300 individuals thus resulting in a minimum average of 25 individuals per sun-sign bin. The X^2 probability obtained for each category in Table 9.2 ranged in value from .07 to 0.84, a result completely in keeping with the assumption of a random distribution. This list could undoubtedly be further expanded if one wished to set forth formal definitions of terms such as "lusty" bones, "swarthy" complexion, "angular" face, "wiry" torso, and the like.

A second commonplace sun-sign correlation in astrology is that which is alleged to exist between a person's career and sun-sign. Examples of the claim are quite commonplace in the astrological literature. As an example, in Table 9.3 we have listed four astrological sources and the careers which each has deemed to be associated with the astrological sun-sign of Leo. As in the case for physical characteristics, there is a certain amount of disagreement in the astrological literature over the particulars of the correlations, but there is no doubt concerning the overall acceptance of the idea. The sun-sign vs. career correlation is perhaps the easiest of the sun-sign correlations to investigate statistically due to the fact that over the past thirty years or so, a wide variety of "Who's Who in . . . " types of biographical collections have been published which contain the birthdates of large numbers of individuals in a given career or occupation. Such collections often contain thousands of biographies and hence provide us with an impressive reservoir of raw data from which to work.

To date, the most complete statistical investigations employing such data are those of Van Deusen (9.7), Culver (9.4) and Culver and Ianna (9.9), which together encompass over sixty separate careers and occupations. In virtually every instance, the calculated X^2 probability rendered the observed sun sign-career distribution statistically indistinguishable from one that was totally random. The results in Table 9.4 are particularly devastating because they not only obliterate the notion of a career-sun sign correlation, but also strongly suggest that personality traits, which are quite often found to be correlated with one's occupation, are also independent of sun-sign.

Direct statistical testing of such a proposition, however, is somewhat difficult. Personality traits, as any social scientist can attest, are essentially abstract in nature and are thus most elusive entities when it comes to attaching numerical values to them. Witness the flap in recent years over the testing of a trait as deceptively simple as one's inherent intelligence via IQ tests. Keeping these inherent difficulties and limitations in mind, there have been a few direct studies (9.10, 9.11, 9.12, and 9.13) conducted on possible correlations between an individual's sun-sign and certain personality traits as measured by several standardized tests. Some of the results are summarized in Table 9.5. The X^2 probabilities once more are never less than 0.05 for any item in Table 9.5 and once more the observed distribution of the given personality trait by sun-sign was found to be statistically consistent with a simple random distribution. In one case (9.13) a curious result was found for a "femininity" index. Persons born in one six month period turned out to differ significantly from those born in the other half of the year. There is no clear reason for this, however the

127

TABLE 9.3   Astrological Listings of "Leonian" Occupations

SOURCE	OCCUPATIONS ASSOCIATED WITH THE SIGN OF LEO
"The Modern Text-Book of Astrology" (Hone); p. 61	Actor, Commissionaire, Chairman, Film Star, Games Professional, Goldsmith, Jeweler, Manager, Monarch, Organizer, Overseer
"The Compleat Astrologer" (Parker and Parker); p. 115	Actor, Dancer, Teacher, Youth Worker, Managing Director, Professional Sportsman, Astrologer, Commissionaire, Jeweler
"Astrological Keywords" (Hall); p. 32	Professional Athletes, Executives, Pioneers, Government Officials, Jewelers, Judges, Money-lenders, Brokers, Foremen, and Electricians
"The Living Zodiac" (Higgins); p. 57	Top Ranking Military Officers, Highly Placed Civil Servants, Diplomats, and Other Government Posts; Professions Allied to High Finance, Stockbroking, and Banking; Principals of Schools and Universities; Legal Professions, Construction, Pioneering, or Work Making Use of Gold

TABLE 9.4   Sixty Occupations Not Correlated to Sun-Sign (9.4, 9.7, 9.9)

OCCUPATION	OCCUPATION
Actors	Industrialists
Advertising Executives	Insurance Agents
Aeronautical Engineers	Journalists
Airline Pilots	Lawyers
Architects	Librarians
Army Officers	Mathematicians
Artists	Mechanical Engineers
Astronomers	Medical Doctors
Atmospheric Scientists	Metallurgical Engineers
Authors	Microbiologists
Bankers	Mining Engineers
Baseball Players	Musicians
Biochemists	Naval Captains
Botanists	Navy Fliers
Business Executives	Oceanographic Scientists
Chemists	Opera Singers
Civil Engineers	Physicists
College Administrators	Poets
College Teachers	Political Scientists
Community Leaders	Politicians
Composers	Popular Singers
Electrical Engineers	Protestant Ministers
Entomologists	Psychiatrists
Film Makers	Psychologists
Football Players	Pugilists
Geologists	Roman Catholic Priests
Geneticists	Sea Captains
Government Officials	Soldiers (United Kingdom)
High School Teachers	Surgeons
Hospital Administrators	Zoologists

TABLE 9.5 Personality Traits Not Correlated to Sun-Sign

Aggression (9.11)	Introversion (9.12, 9.14)
Ambition (9.11)	Intuition (9.11)
Anti-Sociability (9.12)	Knowledge (9.14)
Communality (9.13)	Leadership (9.14)
Creativity (9.11)	Perfection (9.14)
Emotional Maladjustments (9.12)	Power (9.14)
Extroversion (9.11, 9.12, 9.14)	Relationships (9.14)
Feeble-Mindedness (9.12)	Security (9.14)
Flexibility (9.13)	Self-Expression (9.14)
Harmony (9.14)	Sociability (9.14)
Inferiority Feelings (9.12)	Tough-Mindedness (9.14)
Integrity (9.14)	Understanding (9.14)
Intelligence Quotient (9.12)	Wisdom (9.14)

sample was not large and we know of no replication of the study which confirms the effect.

A logical consequence of the astrological belief in a personality-sun sign correlation is the idea that certain combinations of sun-signs (and hence personality types) make for good or "compatible" relationships between individuals while other sun-sign combinations make for bad or "incompatible" relationships. Because of the obvious jump that can be made to astrological love and marriage, listings of compatible and incompatible sun-sign combinations are highly popular items with the public and present one of the most common manifestations of sun-sign astrology. Astrologer Teri King, for example, tells us in "Love, Sex, and Astrology" (9.16) that the marriage of a Scorpio woman with a Virgo man "has far too much against it for it to succeed", while the marriage between a Scorpio woman and a Cancer man is "a splendid union". Not all astrologers share Ms. King's assessments. In fact, there is considerable disagreement in the astrological literature over what combinations constitute "compatible" and "incompatible" sun-sign pairs. To illustrate the magnitude of the disagreement, we have, in Figure 9.1, shaded all of the "incompatible" sun-sign combinations as presented not only by Ms. King but by astrologers Carrol Righter (9.15), Norvell (9.17) and Sydney Omarr (9.18) as well. If we regard divorce as one measure of a couple's incompatibility, then it is possible to statistically resolve this disagreement in sun-sign claims by simply looking at the distribution of the sun-sign combinations that are found to occur in divorces. Such studies have in fact been performed by Silverman (9.19) and Kop and Heuts (9.20) on the respective divorce rates in the state of Michigan and in the city of Amsterdam, Holland. In both studies, the statistical results indicate that none of the sun-sign combinations possessed divorce rates that were significant. In other words, *not one* of the 144 squares making up the sun-sign matrix for actual divorce rates can be shaded with any statistical significance in Figure 9.1.

Far and away the most dangerous claim made for sun-sign astrology concerns the supposed relationship between sun-sign and medical profiles. Basically the idea is that each sun-sign has associated with it a set of physiological strengths and weaknesses and these, in turn, can be used to generate an "astrological map" of the medical problems most likely to arise in a given individual. Table 9.6 lists a few of the astrological summaries of diseases and ailments deemed peculiar to the sun-sign of Capricorn. Again we find the typical astrological concensus on the validity of the general idea but some disagreement on the particulars; and again, statistical investigations of the claims of sun-sign medical astrology turn up blanks.

In his book entitled "Season of Birth", for example, Per Dalen (9.12) summarizes most of the major statistical studies which have dealt with possible correlations between birthdates and various diseases, medical phenomena, and personality traits. A partial listing of these and other results for some of the more familiar diseases and medical phenomena which exhibit *no* correlation with an individual's sun-sign is presented in Table 9.7. These data are of considerable importance since they have been generated by a wide variety of investigators working all over the world. In virtually every case the random results obtained

131

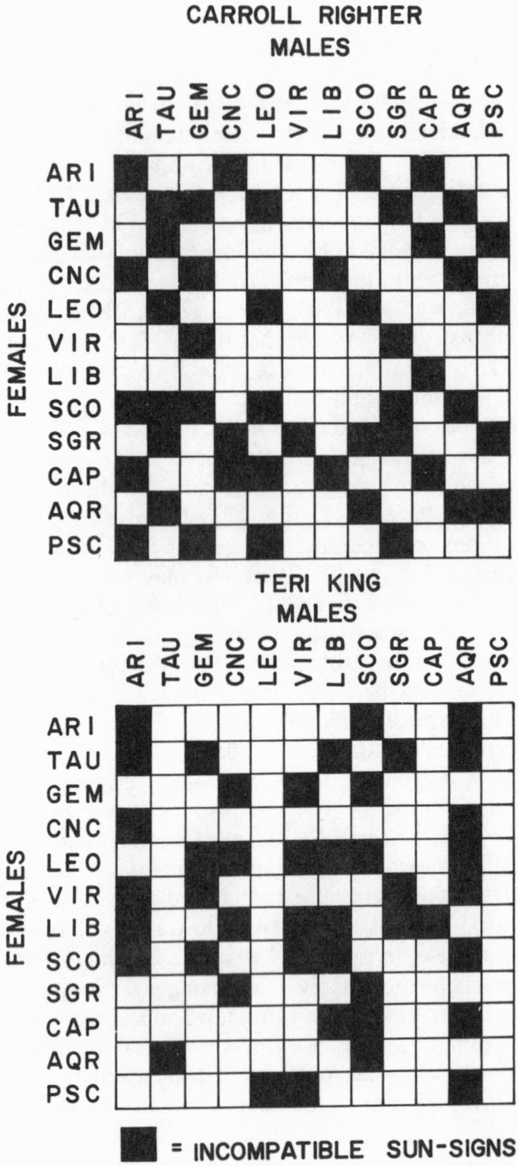

Figure 9.1

Incompatible sun-sign combinations in marriage as published by astrologers Carroll Righter (9.15), Terri King (9.16), Norvell (9.17) and Sydney Omarr (9.18). The observed incompatibilities as measured by actual divorce rates show no statistically significant sun-sign combinations (9.19, 9.20).

132

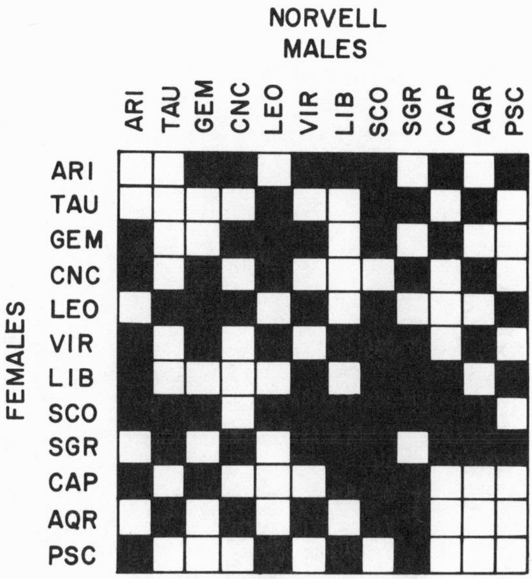

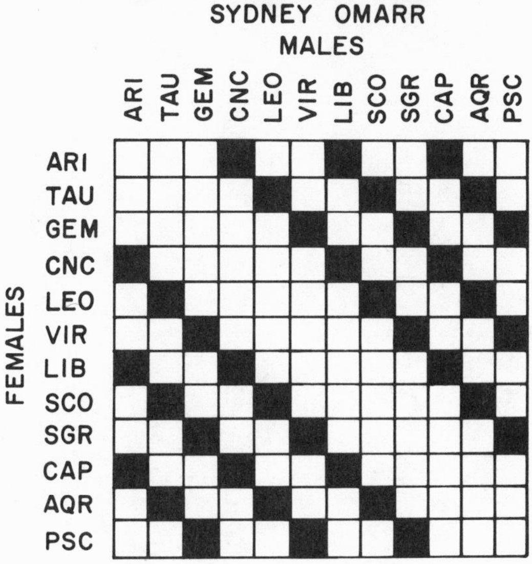

ARI TAU GEM CNC LEO VIR LIB SCO SGR CAP AQR PSC

FEMALES

ARI
TAU
GEM
CNC
LEO
VIR
LIB
SCO
SGR
CAP
AQR
PSC

SYDNEY OMARR
MALES

ARI TAU GEM CNC LEO VIR LIB SCO SGR CAP AQR PSC

FEMALES

ARI
TAU
GEM
CNC
LEO
VIR
LIB
SCO
SGR
CAP
AQR
PSC

■ = INCOMPATIBLE SUN-SIGNS

TABLE 9.6   Astrological Listings of "Capricornian"
Medical Ailments and Diseases

Source	Medical ailments associated with the sign of Capricorn
"Astrology – Your Wheel of Fortune" (Norvell) p. 111	Rheumatism, and arthritis, disorders of the joints of the body; severe colds, influenza, eczema
"Medical Astrology" (Garrison) p. 235	Broken bones, arthritis, deafness, acne, loss of teeth
"The Modern Text-Book of Astrology" (Hone) p. 76	Diseases which limit, such as rheumatism. Orthopaedic troubles; skin troubles
"The Encyclopedia of Astrology" (Shulman) p. 176	Bright's disease, rickets, anaemia, cataracts, boils, deafness, and rheumatism

134

**TABLE 9.7   Some Diseases and Medical Phenomena,
Found To Be Independent of Sun-Sign (9.4, 9.9, 9.12)**

Acne	Leukemia
Allergies	Longevity
Appendicitis	Lung Cancer
Athlete's Foot	Measles
Broken Bones	Moles
Bronchitis	Multiple Births
Chicken Pox	Multiple Sclerosis
Common Cold	Mumps
Congenital Nervous Disorders	Muscular Dystrophy
Constipation	Natural Abortions
Dandruff	Premature Births
Diabetes	Rheumatism
Diarrhea	Rubella
Down's Syndrome	Sprains
Eye Defects (at birth)	Stillbirths
Gout	Strep Throat
Heart Attacks	Strokes
High Blood Pressure	Tonsilitis
Hodgkin's Disease	Torn Ligaments
Infant Mortality	Uterine Bleeding
Influenza	Whooping Cough

are completely at odds with the basic tenets of sun-sign astrology. The only exception to this general trend was for schizophrenia where a significant excess of births in the winter or early months of the year has been reported. Two recent studies, one in the northern (9.21) and one in the southern hemisphere (9.22), have attempted to confirm these findings each using a sample of more than 6000 schizophrenic patients. In both cases no seasonal variations of birth times were found.

The firmly established scientific evidence for the sun-sign's mythical powers has resulted in a most interesting double-tiered astrological response. To the astrologically sophisticated, for example, Sydney Omarr (9.23) tells us

"Of course the Sun-sign meanings are, as we should be well aware of at this state, colored by the House position of the sun, by the aspects, by the fact that other planets are or are not in the same sign occupied by the Sun".

This is the same Sydney Omarr who has written several books on astrology and who currently writes a daily astrological column for the Los Angeles Times syndicate as well as an annual volume entitled "Sydney Omarr's Astrological Guide for You in (year)". All of these dole out astrological advice and prediction by the pageload, not on the basis of all of the factors in a horoscope, but solely on the basis of an individual's *sun-sign*.

In effect then, a large number of astrologers unashamedly work both sides of the street when it comes to the question of sun-sign astrology. On one hand, much of the astrological community trumpets the supposed power of sun-sign astrology via the mass media, only to scurry behind the "other factors present in the horoscope" when confronted with an ever-increasing body of statistical evidence against such astrological sun-sign correlations.

Even sincere attempts by more sophisticated astrologers to correct the situation have run into foul ground with their colleagues. In 1976, for example, nine member associations of the astrologically prestigious Congress of Astrological Organizations denounced sun-sign astrology advice columns in newspapers as "utter nonsense and fraudulent endeavors" (9.24). What impact has this denunciation had on the world of astrology? The daily sun-sign columns continue. The annual astrological sun-sign guides of Sydney Omarr, Carroll Righter, et. al have rolled off the presses exactly on schedule for 1977 and 1978. In fact, in the May 1977 issue of "Horoscope" magazine, astrologer Mark Feldman expressed a good deal of annoyance with his colleagues' denunciation of sun-sign astrology:

"There are a number of reasons why this action on the part of some astrologers and organizations was a mistake. To list a few: the astrological community simply cannot afford to break up its ranks because of an issue as basic to this field as Sun-signs; those astrologers who signed the statement denouncing newspaper columns apparently failed to realize that all astrologers begin with an analysis of the twelve signs of the zodiac; the fact that over a thousand daily newspapers in the United States carry horoscope columns represents the only medium

136

by which the public can be made constantly aware that there are astrologers still working at and devoted to this field; and as Julia Wagner recently commented, "How do you suppose millions of people get interested in astrology?" . . . most of the astrologers who write the newspaper horoscope columns – Sydney Omarr, Charles Jayne, Carroll Righter – are distinguished, with many years of experience in this field. Sun-Sign astrology is not only necessary but also important!" (9.25).

Thus, from a publication that bills itself as "The World's Leading Astrological Magazine", we have the spectacle of sun-sign astrology being justified *not* on the basis of hard experimental evidence, but rather because astrologers can't afford to break ranks, because the sun-sign columns increase public awareness of astrology, and because the people who write them have been at it for years! Unfortunately such sentiments are echoed throughout a large segment of the astrological world. The scientific method meanwhile demands that we follow wherever our experimental results lead us. When corrections, however painful they may be, are required, they are applied, and thus science advances. The fact that much of the astrological community is unwilling to repudiate its cherished sun-sign astrology in spite of the overwhelming amount of statistical evidence against it currently constitutes one of the major stumbling blocks between scientist and astrologer. There are, of course, numerous astrologers who are appalled at this attitude within their ranks. We can only hope their voices will eventually be heard over the din of their colleagues' rush to gain a piece of the highly lucrative sun-sign action.

9.1   Goodman, L., "Linda Goodman's Sun-Signs", p. xviii, Bantam Books, Inc., New York (1968).

9.2   Hone, M., "The Modern Text-Book of Astrology", p. 49, revised edition, L.N. Fowler and Company, Ltd., London (1972).

9.3   Dean, G., and Mather, A., "Recent Advances in Natal Astrology", pp. 83-102, The Astrological Association, Bromley Kent, England (1977).

9.4   Culver, R., "Sun Sign Sunset", Pachart & Publishing House, Tucson (1979).

9.5   Shulman, S., "The Encyclopedia of Astrology", p. 168, Hamlyn Publishing Group Ltd., New York (1976).

9.6   Higgins, T., "The Living Zodiac", p. 33, Black Watch Publishers, New York (1976).

9.7   Van Deusen, E., "Astrogenetics", Doubleday and Company, Inc., Garden City, New York (1976).

9.8   Ibid., ref. 9.4.

9.9   Culver, R., and Ianna, P., *Astron. Qtrly.* 1, 147 (1977).

9.10   Ibid., ref. 9.3, p. 113.

9.11   Silverman, B., and Whitmer, M., "Astrological Indicators of Personality", *J. Psychology*, **87**, 89 (1974).

9.12   Per Dalen, "Season of Birth", American Elsevier Publishing Company, Inc., New York (1975).

9.13   Pellegrini, R., "The Astrological Theory of Personality", *J. Psychology* **85**, 21 (1973).

9.14   Ibid., ref. 9.3, p. 122.

9.15   Righter, C., "Your Astrological Signs Tell If You've Picked the Right Mate", *National Enquirer*, Feb. 8, 1977, p. 49.

9.16   King, T., "Love, Sex, and Astrology", Barnes and Noble Books, Harper and Row, Publishers, New York (1973).

9.17   Norvell, "Astrology – Your Wheel of Fortune", Barnes and Noble Books, Harper and Row, Publishers, New York (1975).

9.18   Omarr, S., "Astrology, You, and Your Love Life", Pyramid Books, New York (1972).

9.19   Silverman, Bernie I., "Contemporary Astronomy" by J. Pasachoff, cf p. 437, W.B. Saunders Company, Philadelphia (1977).

9.20   Kop, P. and Heuts, B., *J. Interdiscipl. Cycle Res.* **5**, 19 (1974).

9.21   Koehler, K., and Jacoby, C., *Archiv fur Psych. und Nervenkr.*, **223**, 69 (1976).

9.22   Krupinski, J., Stoller, A., and King, D., *Austr. and NZ Jour. Psych.*, **10**, 311 (1976).

9.23   Omarr, S., "My World of Astrology", p. 195, Wilshire Book Company, North Hollywood, California (1975).

9.24   *The Humanist*, Sept./Oct. 1976, p. 66.

9.25   Feldman, M., *Horoscope Magazine*, May 1977, pp. 41-42.

Chapter 10

Testing the Astrological Gestalt

The late Senator Hubert Humphrey once commented in the 1960 West Virginia primary that campaigning against the Kennedys was like campaigning against a centipede; there were so many arms and legs to deal with. So it is with astrology. As one leaves the relative simplicity of sun-sign astrology one finds that the astrological edifice fragments into a multitude of astrological feudal states, each with its own set of rules, correspondences, and synchronicities. Any attempts to test correspondences are thus akin to dealing with the astrological equivalent of the good Senator's centipede.

We have already described the lack of astrological agreement over some of the basic issues involved in the casting of a horoscope, such as aspects, orbs, house divisions, and sidereal versus tropical zodiacs. It is in the realm of the rules of interpreting the horoscope, however, that the "entropy monster" enjoys its finest astrological hour. For example, if the planet Mercury is in the astrological sign of Cancer, Sydney Omarr (10.1) tells us:

"Mercury in Cancer is described by a major keyword: adaptability."

Mark Edmund Jones (10.2) disagrees:

"Mercury in Cancer ... means a mentality which is culminative, dogmatic, and circumscribed, rather than original, adjustable inquiring."

In the face of this type of diversity of interpretation, one cannot hope to test each and every astrological claim made for each and every planetary array that appears in the horoscope. What can, and in some measure, has been done, is to test an astrological array to see if *any* configurations are statistically significant.

One of the easier astrological alignments to test is the location by astrological

139

TABLE 10.1   The Motion of Mars Through
The Astrological Signs In 1970-71

Sign	Dates In Sign	Total Days Spent In Sign	Fraction Of Time Interval Spent In Sign
Aries	Jan 25, 1970 – March 7, 1970	42	.060
Taurus	March 8, 1970 – April 18, 1970	42	.060
Gemini	April 19, 1970 – May 31, 1970	43	.061
Cancer	June 1, 1970 – July 18, 1970	48	.068
Leo	July 19, 1970 – Sept 3, 1970	47	.067
Virgo	Sept 4, 1970 – Oct 20, 1970	47	.067
Libra	Oct 21, 1970 – Dec 6, 1970	47	.067
Scorpio	Dec 7, 1970 – Jan 23, 1971	48	.068
Sagittarius	Jan 24, 1971 – March 13, 1971	49	.070
Capricorn	March 14, 1971 – May 5, 1971	53	.076
Aquarius	May 6, 1971 – Nov 6, 1971	185	.264
Pisces	Nov 7, 1971 – Dec 26, 1971	50	.071

sign of a given planet at the time of birth. The statistical analysis proceeds in essentially the same fashion as for the sun-signs. A given characteristics of an individual is chosen, such as personality traits, etc. and a reasonably large sample of such people is then sorted according to the astrological sign in which the planet of interest was located at birth (Mercury in Cancer, etc.). However we cannot necessarily assume, as in the case for the sun-sign, that the expected distribution of a planet's position by astrological sign is a random one. For example, as the planet Mars moved about the sun in the time period January 25, 1970 when it entered the sign of Aries until December 27, 1971 when it once more entered that sign, it did not spend an equal amount of time in each sign as viewed from earth. In fact from the data in Table 10.1, we find that Mars spent over ¼ of the entire time period in a single astrological sign — Aquarius. This effect is due to the fact that the earth was catching up to and passing Mars during this time period. As a result of this relative motion, Mars appears from earth to stop its normal west to east motion, back up for a while, stop again, and then resume its normal easterly motion. The net result of this backward or "retrograde" motion was that Mars spent a disproportionate amount of time "fooling around" in the sign of Aquarius.

Thus if we wish to examine a sample of birth dates which were randomly distributed over the interval January 25, 1970 to December 26, 1971 we would *expect* that about ¼ of those birthdates would show Mars in the sign of Aquarius if there were no other correlations, simply because Mars was in Aquarius for ¼ of the time interval. Therefore, in making our X^2 comparison for a planet position we must replace the assumption of a random distribution with one which reflects the amount of time the given planet spent in each sign during the total time interval involved. Once we make this change, we are then ready to proceed with a X^2 search for possible correlations between human attributes and the sign in which a given planet was located at birth.

Several such studies have been carried out for a number of occupations as well as personality, physical, and medical traits. In each case the birth data were sorted according to the sign in which the planets Mercury, Venus, Mars, and Jupiter were located at the time of birth. One study, that of the Gauquelins (10.3), also included the moon-sign as well as the sign in which Saturn was located at the time of birth. In every instance tested, not one X^2 probability was less than 0.08, thus indicating that not one statistically significant departure occurred from the assumed distribution of planetary positions over the time intervals involved. A total of twenty-three occupations, thirteen medical problems, two physical characteristics, and one personality characteristic were tested in this fashion. (See Table 10.2)

One can also use the data to check the locations by astrological sign of any two planets, e.g. does the Mars in Scorpio-Venus in Virgo combination occur significantly more often than other astrological sign combinations of the same two planets. The problem here, however, is that the ante immediately goes from 12 to 144 possible combinations, since each of the two planets to be tested can be independently located in any one of the twelve signs. This increase in turn requires a much larger sample size if the method is to be effective. If we wish to have a minimum average of 25 counts per possible planet-astrological sign

141

**TABLE 10.2   Human Qualities Not Correlated To The Sign At Birth
In Which Mercury, Venus, Mars, or Jupiter Was Located* (10.3, 10.4, 10.5)**

OCCUPATIONS

Actors
Athletes
Bankers
Baseball Players
Cabinet Ministers
College Teachers
Engineers
Executives
Journalists
Lawyers
Medical Doctors
Ministers
Musicians
Painters
Playwrights
Poets
Politicians
Psychologists
Scientists
Soldiers
Solo Athletes
Team Athletes
Writers

MEDICAL PROBLEMS

Alcoholism
Appendicitis
Asthma
Chicken Pox
Common Cold
Heart Attacks
Malaria
Mumps
Muscular Dystrophy
Polio
Rheumatism
Strokes
Tonsilitis

PHYSICAL CHARACTERISTICS

Height
Longevity

PERSONALITY CHARACTERISTICS

Intelligence Quotient

*The Gauquelins' (10.3) results include null results for the Moon and the
planet Saturn as well.

142

combination, we would thus need a sample size in the present case of at least 3600. Most of our sample sizes relating to occupations do indeed satisfy this criterion, but unfortunately, the sample sizes relating to the medical, personality and physical characteristics were too small to be employed. We also attempted to limit the planet combinations tested to those planets which astrologers seemed to most associate with the given occupation (10.6). In each case the resulting distribution was found not to be significant and no statistically significant correlations were evident. (Table 10.3).

Statistical testing for significant aspects or angular separations can be performed in much the same way as above. The assumed distribution of angular separations in this case is based on the observed mean fraction of the planet's synodic period spent in each angular separation interval. On the basis of such a distribution for Mercury and the Sun, for example, we would expect Mercury to be within a 6° orb of the Sun about 12% of the time (see Table 10.4-I). Distributions of angular separations were calculated for the following planet combinations: Sun-Mercury, Sun-Venus, Sun-Mars, Sun-Jupiter, and Sun-Saturn. The distribution finally used for each planet was the composite of the distributions of 25 synodic periods* of the planet. These angular separation distributions were then in turn compared with the observed angular separations of the same planet pair for the items listed in Table 10.5. No statistically significant departures from the assumed angular separation distribution were found for any of the items in Table 10.5.

In the foregoing sections we have made use of and referred to statistical analyses of sets of birthdates and their corresponding planetary configurations in an attempt to verify at least in principle the validity of horoscope analysis. Unfortunately the data employed are birthdates which are given by day, month, and year, but *not* by exact time of day the birth occurred. Without such information, as we have seen, astrologers are somewhat reluctant to erect and interpret a chart, since access to important astrological information such as the location relative to the zodiac of the astrological houses and certain astrologically key points including the ascendent, midheaven, etc. would be effectively denied to them. In a single 24 hour period, for example, the ascendent point can be found at any point along the zodiac, depending on the relative orientation of the horizon and the ecliptic.

To date the only set of birthdate data having a reasonable size and which includes the time of day of the birth is that collected by Michael and Francois Gauquelin (10.3). Over a period of several years, the Gauquelins assembled and analyzed the complete horoscopes of some 25,000 individuals. These data have allowed Michael Gauquelin to investigate in detail some of the astrological claims which rest, not only on the date of birth, but on the time of day as well.

For example, the presence of Jupiter in conjunction with an individual's medheaven at birth:

*A synodic period is the time it takes the earth to gain or lose one full "lap" relative to another planet's orbital motion around the sun, and hence is the time required for a planet to return to the same aspect relative to the sun.

**TABLE 10.3   Two-Planet Combinations By Astrological Signs
At Birth Not Correlated To Occupation**

OCCUPATION	PLANET COMBINATIONS TESTED
Army Officers	Mars-Sun, Mars-Mercury, Mars-Venus
Bankers	Mercury-Sun, Mercury-Venus, Mercury-Mars
Baseball Players	Mars-Sun, Mars-Mercury, Mars-Venus
Engineers	Mercury-Sun, Mercury-Venus, Mercury-Mars
Musicians	Venus-Sun, Venus-Mercury, Venus-Mars
Poets	Venus-Sun, Venus-Mercury, Venus-Mars
Politicians	Sun-Mercury, Sun-Venus, Sun-Mars
Psychologists	Sun-Mercury, Sun-Venus, Sun-Mars

TABLE 10.4 The Distribution Of Angular Separations Of (I) The Sun and Mercury and (II) The Sun and Jupiter

| I | | II | |
| SUN-MERCURY | | SUN-JUPITER | |
Angular Separation Interval	Mean Fraction Of Synodic Period Spent In Interval*	Angular Separation Interval	Mean Fraction Of Synodic Period In Interval*
East 0°– 3°	.029	0°– 15°	.051
3 – 6	.033	15 – 30	.050
6 – 9	.037	30 – 45	.049
9 – 12	.042	45 – 60	.048
12 – 15	.048	60 – 75	.047
15 – 18	.063	75 – 90	.045
18 – 21	.081		
21 – 24	.098	90°– 105°	.042
24 – 27	.070	105 – 120	.038
		120 – 135	.035
West 0°– 3°	.029	135 – 150	.033
3 – 6	.033	150 – 165	.032
6 – 9	.037	165 – 180	.031
9 – 12	.042		
12 – 15	.048	180°–195°	.031
15 – 18	.063	195 – 210	.032
18 – 21	.081	210 – 225	.033
21 – 24	.098	225 – 240	.035
24 – 27	.070	240 – 255	.038
		255 – 270	.042
		270°–285°	.045
		285 – 300	.047
		300 – 315	.048
		315 – 330	.049
		330 – 345	.050
		345 – 360	.051

*Based on an average of 25 synodic periods

145

TABLE 10.5   Items Independent Of The Angular Separation At Birth
Of The Following Planetary Pairs:
Mercury-Sun, Venus-Sun, Mars-Sun, Jupiter-Sun, Saturn-Sun

OCCUPATIONS	MEDICAL ITEMS:
Army Officers	Athletes' Foot
Bankers	Chicken Pox
Baseball Players	Common Cold
Chemists	Heart Attacks
Engineers	Influenza
Musicians	Longevity ($\geqslant$ 80 years)
Physicists	Mumps
Poets	Strep Throat
Politicians	Strokes
Psychologists	Tooth Decay

" . . . is an excellent aspect for success in life, giving the right positive attitudes towards both the career and general objectives in life." (10.7).

Out of ten thousand successful individuals examined in one aspect of the Gauquelins' survey, the percentage of individuals who had Jupiter in conjunction with the midheaven point at birth was not different from what would be expected for pure chance.

The planet Mars has long been astrologically associated with blood and violence. Shulman (10.8) for example tells us that Mars:

" . . . represents violence and sexuality . . . gives a masculine, aggressive, and fearless nature which can be overbearing to the point of tyranny."

Hall (10.9) sees some of the Martian characteristics as:

" . . . Destructive, passionate, egotistic, coarse, sarcastic, ironical, cruel, treacherous, quarrelsome, warlike."

The Gauquelins proceeded to assemble the files of 623 murderers who, in Michael Gauquelin's words, "according to the judgment of experts, were the most notorious in the annals of justice for the horror of their crimes" (10.10). The distribution of Mars' location by astrological house was then compared with a demographic distribution of birthdates and no statistically significant correlations were found, even for the eighth house of "death for oneself or for others" or for the twelfth house, which rules over "trials and goals".

In fact, in all of the Gauquelin analyses, the only statistically meaningful correlations which could be developed were of a diurnal nature. The Gauquelins claim that their data lead to statistically significant correlations between certain occupations and a certain planet's hour angle* at the time of birth. These results are summarized in Table 10.6 and have sparked not only a complete theoretical explanation by Gauquelin himself, but also an acrid debate on the topic, the major points of which can be found detailed in a series of articles by the Gauquelins and L.E. Jerome in *Leonardo*, Volumes 7-9. Basically, the Gauquelins' analysis has been criticized on a number of counts, the bulk of which center on the counting scheme and the computation of the number of degrees of freedom in the data. The claim is made that if these factors are properly introduced, the Gauquelins' odds against a random distribution drop considerably from the value of roughly 10^{-6} quoted in their early work on the subject.

In particular, one analysis of the Gauquelins' so-called "Mars effect"** by Abell, et al (10.12) has failed to statistically confirm the effect. The authors, however, stress that while their findings can place an upper limit on the total strength of the effect, they cannot in any way be interpreted as disproving its

*The object's angular distance east or west of the celestial meridian (north-south line).

**An aspect of the Gauquelins' data in which a statistically significant number of sports champions are found to be born when the planet Mars is situated between the eastern horizon and the celestial meridian.

TABLE 10.6   A Summary Of Gauquelin's Planetary Correlations (10.11)

RISE AND/OR MERIDIAN OF	HIGH BIRTH FREQUENCY	AVERAGE BIRTH FREQUENCY	LOW BIRTH FREQUENCY
Moon	Ministers Politicians Writers	Scientists Doctors Painters Musicians Journalists	Athletes Soldiers
Mars	Scientists Doctors Athletes Soldiers Executives	Cabinet Ministers Actors Journalists	Writers Painters Musicians
Jupiter	Team Athletes Soldiers Ministers Actors Journalists Playwrights	Painters Musicians Writers	Solo Athletes Scientists Doctors
Saturn	Scientists Doctors	Soldiers Ministers	Actors Painters Journalists Writers

existence.

The Gauquelins' data are, however, generally regarded as being sufficiently impressive to at least warrant additional testing. Unfortunately, the debate over the Gauquelins has, to date, centered on the interpretation of one set of data (the Gauquelins') collected by one investigative team (also the Gauquelins) in one part of the world (Europe). It would, for example, be highly interesting to see if exactly the same effect holds in other parts of the world such as the United States, etc. Unfortunately birth records outside of Europe are quite often fail to specify the time of day at which the birth occurred, thus making it very difficult to assemble the data needed to verify the existence of these correlations elsewhere in the world. Thus the question of whether the surprising results achieved by the Gauquelins are due to peculiarities in the data or to actual correlations with planetary positions remains unresolved.

Of all the valid statistical analyses done on the alleged astrological correspondences, the results displayed in Table 10.6 represent to date virtually the only direct statistical aid and comfort to the world of the astrologer. It is not surprising then, that the astrologers clasp this one aspect of the Gauquelins' findings to their hearts with a fervor normally reserved for a talisman.

A typical astrological reaction to the Gauquelins' work, for example, is the commentary of West and Toonder:

"But perhaps the same tenacity that permits Gauquelin to press on ... also prevents him from admitting or even seeing, that he is proving astrology, plain old-fashioned astrology." (10.13)

Astrologer Francis King bubbles even more optimistically:

" ... his (Gauquelin's) results showed that the classic astrological theories appeared to have been scientifically vindicated." (10.14)

Even a cursory look at Michael Gauquelin's published work clearly indicates that to place such interpretations on it constitutes a gross misrepresentation of these results. The Gauquelins have demonstrated as thoroughly as any investigator to date that, in Michael Gauquelin's own words:

"It is now quite certain that the signs in the sky which presided over our births have *no power whatever* (italics ours) to decide our fates, to affect our hereditary characteristics, or to play any part, however humble, in the totality of effects, random or otherwise, which form the fabric of our lives and mold our impulses to actions." (10.15)

And in "Cosmic Clocks", Gauquelin continues the assault:

"Every effort made by astrologers to defend their basic postulate, that the movement of the stars can predict destiny has failed ... Statistics have disposed of old arguments once and for all: the numbers speak without bias, and they leave no room for doubt. Whoever claims to predict the future by consulting the stars is fooling either himself or someone else." (10.16)

The Gauquelins' work thus provides us with yet another fine example of the double vision with which a great way astrologers are afflicted when it comes to a discussion of empirical results relating to astrology. On one hand, the very small

149

percentage of the Gauquelins' results which suggest that certain celestial correspondences may exist are widely trumpeted as "proof" of the validity of *all* astrology. Meanwhile all of the null results obtained by the Gauquelins for a most impressive number of supposed astrological correlations and correspondences, using the very same data set and statistical methods, are either conveniently ignored by astrologers or are dismissed with arguments not dissimilar to the following:

"... he (Gauquelin) commits some of the same errors of over-simplification by refuting various factors singly. Thus he considers signs alone, aspects alone, houses alone, etc., and finds no meaning in them. But the one primary rule in astrology is that no factor can be taken out of context without a real danger of losing the meaningful gestalt." (10.17)

Scientists cannot help but marvel at a "meaningful gestalt" which seems to be lost only when the testing of individual factors in a horoscope yields null results, although a great many experimental and observational scientists have undoubtedly long yearned for the existence of such behavior in their own areas of endeavor!

The concept of the astrological gestalt, however, does raise some very serious questions concerning our very ability to test astrological correspondences. If, for example, we wish to statistically analyze the distribution at birth not of one but three planets by astrological sign, we would have to deal with a total of $(12)^3$ or 1728 possible combinations of planets and astrological signs. Four planets can take on over 20,000 possible combinations with astrological signs, and if we consider all ten astrological planets, there are $(12)^{10}$ or about 60 billion possibilities involving just the distribution of the planets by astrological sign at the time of birth. Such numbers simply overwhelm any statistical samples that could ever be assembled. Even a four planet test of the Gauquelins' sample of 25,000 individuals, for example, would have the totally undesirable statistical effect of scattering this impressive sample into an average density of barely one person/combination bin. And if we were to even require this unsatisfactory average statistical density for a combination of all ten astrological planets, we would need a sample size comparable in number to the sum total of every human being who has ever lived on this planet! Keep in mind also, of course, that the astrological gestalt is still incomplete, since we have yet to include aspects, houses, ascendents, etc.*

But if the sheer enormity of the astrological gestalt creates virtually insurmountable problems for statistical analyses, it again raises potentially fatal questions regarding the veracity of the "wisdom of the ages" from whence these astrological correspondences alledgedly sprang. At the time of Ptolemy, for instance, one can estimate (10.18) that the total world population was about 50

---

*A complete test of the astrological gestalt would involve at least 10^{35} possible combinations of astrological points, planets, signs, houses, and aspects! By contrast an estimate of the total number of grains of sand on all the world's beaches is of the order of 10^{27} grains.

million. Thus, an astrologer with access to the horoscope of *every* person on the planet in 150 AD could conduct a meaningful statistical test of the distribution of no more than six planets by astrological sign at birth. Correspndences involving various astrological points, houses, and aspects would once more go untouched. We can once more conclude from such data that the astrological correspondences handed down through the centuries have been developed not on the basis of the entire horoscopic package, but rather *within* the horoscope, using the very isolated correspondences whose wreckage can be found strewn throughout the numerous statistical investigations of the past several decades.

10.1   Omarr, S., "My World of Astrology", p. 248, Wilshire Book Company, North Hollywood, California (1975).

10.2   Jones, M., "Astrology — How and Why it Works", p. 295, Shambhala Publications, Inc., Boulder, Colorado (1969).

10.3   Gauquelin, M., "L'Influence des Astres, Etude Critique et Experimentale", Dauphin Press, Paris (1955).

10.4   Culver, R., and Ianna, P., *Astron. Qtrly.*, 1, 85 (1977).

10.5   Barth, J., and Bennet, J., *Leonardo*, 7, 235 (1974).

10.6   See for example: Hall, M., "Astrological Keywords", pp. 118-119, Littlefield, Adams and Co., Totowa, New Jersey (1958).

10.7   Ibid., Ref. 10.2, p. 146.

10.8   Shulman, S., "The Encyclopedia of Astrology", pp. 182-183, Hamlyn Publishing Group Ltd., New York (1976).

10.9   Ibid., Ref. 10.6, p. 112.

10.10   Gauquelin, M., "The Cosmic Clocks", p. 84, Henry Regnery Company, Chicago (1967).

10.11   Gauquelin, M., "The Scientific Basis for Astrology", p. 164, Stein and Day Publishers, New York (1969).

10.12   Abell, G., Abell, D., Gauquelin, M., and Gauquelin, F., *The Humanist*, Sept./Oct. 1976, p. 40.

10.13   West, J. and Toonder, J., "The Case for Astrology", p. 172, Penguin Books Inc., Baltimore, Maryland (1973).

10.14   King, F., "The Cosmic Influence", p. 41, Doubleday and Company, Inc., Garden City, New York (1976).

10.15   Ibid., Ref. 10.11, p. 145.

10.16   Ibid., Ref. 10.10, pp. 85-86.

10.17   Dobyns, Z., *Psychology Today*, September, 1974, p. 131.

10.18   Enrlich, P., "The Population Bomb", p. 18, Ballantine Books, Inc., New York (1968).

Chapter 11

Astrology With a Short Deck

Throughout the span of human history, few numbers have enjoyed the magical significance that has been conferred on the number seven (11.1). Among the dozens of famous groups of seven that dot human lore and mythology we find the seven elders who assisted the Babylonian hero Marduk, the seven Pleiades sisters of Greek mythology, and the seven Wise Ones from the lore of ancient Egypt (11.2). And indeed, one need only to get into a crap game in order to observe one of the stronger aspects of the mystical significance of the number seven in modern society! The origins for this attachment to the number seven are lost in the shadows of prehistory, but the best explanation is that the number's significance is derived directly from the fact that in ancient times seven celestial objects, the five naked eye planets and the sun and moon, were identified as having the ability to move across the sky relative to their stationary stellar counterparts. To the ancients, then, these "Seven Powers" were the watchmen of the heavens who wandered among the stars and could influence terrestrial events. Even to this day these objects are still denoted by their deistic namesakes such as Mercury, Venus, Mars, and so on.

From the time of Ptolemy until well into the 18th century, Western astrologers had almost totally embraced the so-called "Law of Seven" which was, according to West and Toonder:

> " ... held sacred as the sum of spirit and matter* and the convenient septenary of five planets, sun, and moon made a splendid physical 'proof' of the sacred nature of the law ... " (11.3)

*"Spiritual" qualities include fixed, mutable, and cardinal; "Material elements" are earth, fire, water, and air.

152

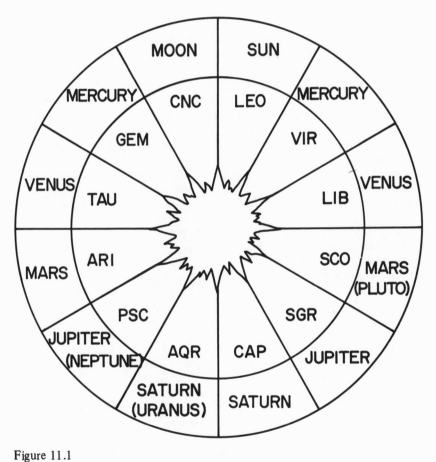

Figure 11.1

The Planetary Relationships in the Sacred Law of Seven. The rulerships of
Uranus, Neptune, and Pluto are also shown.

Further support was lent to this idea by the fact that the planetary rulerships of the astrological signs could also be fit into the scheme with relative ease. Thus, the sun and moon ruled one sign each, while the ruling deities of the remaining five planets were split over two signs each (see Figure 11.1). Such was the state of affairs in Western astrology when, on the night of March 13, 1781 the guns of the Scientific Revolution drew nigh.

Sir William Herschel had come to England as a musician in 1757 and by the late 1770's had developed a deep interest in making telescopes and astronomical observations. In August of 1779, he had begun the ambitious task of observing all of the stars in the night sky down to 8th magnitude* in an effort to locate and catalogue double stars. On that fateful evening of March 13, 1781, Herschel made the following entry in his observing log:

"Tuesday, March 13 (1781). In the quartile near ζ Tau the lowest of the two is a curious either nebulous star or perhaps a comet. A small star follows the comet at 2/3 of the field's distance." (11.4)

At first Herschel thought he had discovered a new comet, and in fact reported it as such to the Royal Society. But as the weeks passed, it became obvious that Herschel's find was not a comet, but rather a full-fledged planet swinging about the sun in an almost perfectly circular orbit with a radius of over 19 astronomical units or nearly three billion kilometers. In one night, Herschel had extended the boundaries of the solar system more than one billion kilometers!

A number of suggestions were offered for the new planet's name, including, among others, Georgium Sidus (Georgian Planet, after King George III of England), Herschel, and Cybile. The name that finally prevailed was Uranus, who in Greek mythology is the father of Saturn. Thus, the ancient tradition of naming planets for gods and goddesses was retained.

The outer planet odyssey, however, had only begun for the astronomical world. As astronomers watched Uranus plod along its 84 year long cycle of the heavens, revisions of its preliminary orbit were computed so that more accurate predictions could then be made for its future celestial positions. Much to their surprise, however, astronomers found that instead of improving their predictive accuracy, the revised orbits led to serious discrepancies between the observed and predicted positions of Uranus. At first, the errors were thought to be due to the small size of the observational arc to be fit by the orbital calculations. Indeed in 1781, the Finnish mathematician Anders Lexell had even fit a parabolic orbit to the planet's observed path in which the perihelion distance was 16 astronomical units and the time of perihelion passage was to be April 10, 1789. Throughout the rest of the century, as Herschel's planet extended the size of its observed arc about the sun, however, it was clear that the orbital arc was not the source of the astronomers' troubles.

Nor was the discrepancy instrumental in nature. By 1840, after some 60 percent of its orbital path had been traversed under the watchful eyes of the

*A brightness roughly 6 times fainter than the faintest stars visible to the naked eye.

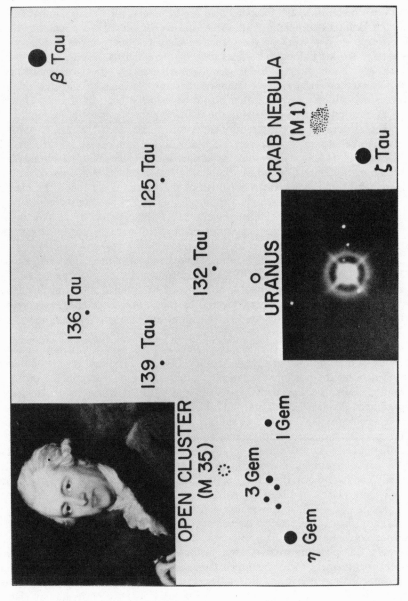

Figure 11.2   Sir William Herschel, Uranus, and the Gemini-Taurus star field on the night of March 13, 1781.

world's astronomers, the discrepancy between the observed and predicted positions of Uranus amounted to roughly one arcminute or 1/60 of a degree, an angle detectable with the unaided eye. By contrast, the telescopes of the day could easily measure positions to an accuracy of 1/600 of a degree!

If the observations were not to blame, then clearly something was wrong elsewhere. After enjoying a century and a half of stunning triumphs using Newtonian mechanics, astronomers were loathe to cast the Newtonian baby out with the Uranus discrepancy bathwater, and so considered the possibility that other forces were at work on the planet which had not been taken into account previously. Working along these lines, two mathematicians, Urbain Leverrier of France and John Adams of England, simultaneously, but independently, explored mathematically the possibility that the discrepancy was due to the gravitational effects of yet another planet beyond the orbit of Uranus. Out of these calculations came a position in the sky where the supposed planet could be seen. The difficulties encountered by these men, particularly Adams, in getting an astronomer to make the suitable searches for the planet are stories all in themselves (11.5, 11.6), but Leverrier was finally able to obtain the cooperation of Johann Galle of the Berlin observatory in the hunt. Armed with a large telescope and an excellent star map of the region, Galle found the planet on the very first night of search within a degree of the positions predicted by both Adams and Leverrier! (See Figure 11.3) Once more, in keeping with tradition, the new planet was named for a deity. This time Neptune, the Roman god of the sea, was to be the god so honored.

The predictive discovery of Neptune gave Newtonian mechanics one of its finest hours. Two individuals using only their wits and their pencils successfully predicted from Newtonian theory, not only the existence of a trans-Uranian planet, but almost exactly where it could be found in the heavens. Leverrier, in fact, never once laid eyes on the planet whose existence and position he had so successfully deduced.

With the discovery of two major planets as well as a number of smaller asteroids, within the span of a single century, it was only natural for astronomers to speculate on the possible existence of additional worlds beyond the orbit of Neptune. Thus, in the early years of the 20th century, two American astronomers, Percival Lowell and W.H. Pickering, published predicted orbits for the "trans-Neptunian" planet based on what they perceived as anomalous behavior in the motion of both Uranus and Neptune.

Unfortunately, Lowell's death in 1916 deprived the astronomical community of the trans-Neptunian planet's most ardent supporter, and as a result, the searches conducted for the object were largely sporadic until 1929, when the Lowell Observatory in Flagstaff, Arizona placed a 13-inch refracting telescope into service. By 1929, it was recognized that if the planet did indeed exist, it would not display an easily discernable disk. Thus, the only way that its presence could be ascertained, was from its motion relative to the background starfields. Complicating matters even further, was the fact that as one goes to fainter brightnesses, there are much larger numbers of field stars to contend with. In response to these difficulties, astronomers developed an ingenious method by which the planet could be sought out. First, two photographic plates

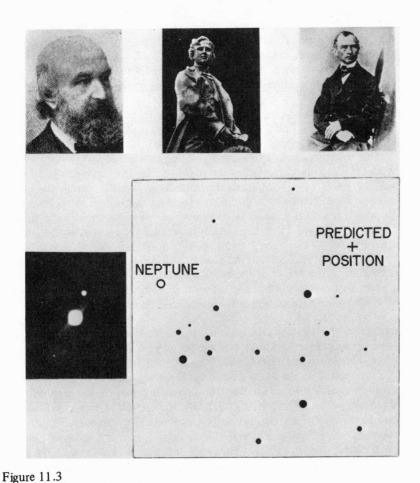

Figure 11.3

Top: The *Dramatis Personae* of Neptune's Discovery. (l to r) Adams, Leverrier, and Galle. Bottom: Neptune and the Aquarius star field on the night of September 23, 1846. The cross indicates the position predicted by Leverrier.

were taken of the same region of the sky several nights apart. The two plates were then optically superimposed in an instrument called a "blink" microscope. By using a shutter which could alternately cut off the illuminating light for one plate and then the other, any image that had moved during the time interval between the two plate exposures would thus appear to "jump" back and forth in the microscope viewer. In this way, a single observer could easily check literally thousands of plate images for any apparent motion in a relatively short time. Thus, a systematic photographic search for the predicted planet was begun at the Lowell Observatory in September, 1929 by a young astronomer named Clyde Tombaugh. After scanning several million star images by the above blink comparison technique, Tombaugh finally found the new planet in February, 1930 within six degrees of the position predicted by Lowell fifteen years earlier. The following month, March 1930, Tombaugh's discovery was announced to the world and the new planet was named Pluto for the Roman god of the underworld.

Although the planet's position was very close to that predicted by Lowell, astronomers have also found that Pluto is much smaller and less massive than Lowell's hypothetical planet ought to be, and, to this day, astronomers have been unable to satisfactorily explain this seeming contradiction.

The planet Pluto marks the last of the so-called "modern" planets. Searches of one sort or other have been made for additional worlds, but in the nearly half-century since Tombaugh's discovery, the best that astronomers have been able to come up with is a rather large number of asteroids, some of which, such as Chiron (see Chapter 5), exhibit highly interesting properties as they wheel about the sun.

In each of these astronomical discoveries, we have witnessed a magnificent demonstration of the power of the techniques and instrumentation with which the modern astronomer probes the heavens. In each of these same discoveries, we have also witnessed a similarly magnificent demonstration of a glaring failure of horoscopic astrology.

Throughout the astrological literature we are constantly reminded of the delicate and interdependent nature of the horoscope. The Astrological Gestalt, so we are told, cannot be tested reliably on isolated correspondences alone, but must be interpreted on the basis of *all* of the factors present in the horoscope by an astrologer qualified to do so. The question thus posed by the discoveries of Uranus, Neptune, and Pluto is why did not astrologers, at some point in the dozens of centuries of astrological history, ascertain the presences of these three planetary influences? After all these objects do add up to a horoscopic presence of some *thirty percent* of the planetary influences. Where were Ptolemy, Nostradamus, William Lilly, John Dee, Madame Balvatsky, Evangeline Adams, and all of the other astrologers whose pre-Plutonian achievements are proclaimed far and wide in the astrological literature? Could not even one of these storied astrologers have ascertained that they were conducting their affairs with an astrologically short deck? The influences of Uranus, Neptune, and Pluto surely *must* have flashed like astrological beacons in the night each and every time they were aspected by one of their less distant naked eye planetary brethren (see Figure 11.5), yet, nowhere in Western astrology do we find any overwhelming

Figure 11.4

Percival Lowell (top left), Clyde Tombaugh and two plates of the planet Pluto, showing its motion among the stars over a several day period.

desire to depart from the Sacred Law of Seven prior to 1781. In fact, Shulman tells us that even after the discovery of Uranus:

"At first astrologers ignored this new arrival (Uranus) and stuck grimly to the planets everyone knew." (11.7).

Incredibly, other astrologers argued at the time that because the new planets could not be seen with the naked eye,* they could not influence human affairs (11.8).

In recent times, there has been a somewhat higher level of sophistication in the astrological arguments advanced to account for this predictive failure. Currently, the most popular concept is the idea that the new planets are "generation" planets whose influence:

"applies to 'the times,' the generations, rather than to individual characteristics ... it is absolutely necessary to remember that Pluto, Neptune, and Uranus ... depict the tenor of the times, the characteristics of the generations, the mood, the society, the tempo." (11.9).

This "generational" influence is presumed to arise from the relatively large amounts of time each planet spends in a given 30° astrological sign.*

Along similar lines, other astrologers claim that:

"The new planets do not manifest themselves in the consciousness of individuals in humanity at large, except in a few outstandingly gifted or 'developed' individuals." (11.10).

One particularly interesting line of defense on this issue comes from astrologers who claim that the astrological influences of any new planets do not begin until the first instant of their discovery! (11.11).

Such viewpoints, however, are a bit difficult to reconcile with the fact that, as we have seen earlier, the aspects of the other planets relative to Uranus, Neptune, and Pluto are almost universally deemed to have astrological importance for the horoscopes of *individual* human beings.

In fact, while hammering away at a horoscope cast by an astronomer in which the planet Pluto was inadvertently omitted, Sydney Omarr comments:

"Anyone attempting to debunk astrology should at least be well enough informed to know the planets. He should know better than to leave out the planet Pluto. Not even a *neophyte* (italics ours) would do that." (11.12).

Since astronomers do not make their living by casting horoscopes, they can perhaps be excused in part for their errors. We will, however, let Mr. Omarr's own words summarize the pre-1930 performance of his astrological predecessors.

The astrological community, meanwhile, has bravely tried to internally adjust

*Actually Uranus at 6th magnitude can be just viewed without optical aid.

*7.0 years for Uranus, 13.7 years for Neptune and 20.7 years for Pluto.

160

to these new worlds. We have already seen for example, that each member of this planetary trio now enjoys a full complement of astrological properties and keywords. In spite of three direct hits in less than two centuries, the grand old Sacred Law of Seven has incurred only minor damage. Astrologers have simply started another "octave" of planets, much as one fills up rows of chemical elements in the periodic table. Thus, Uranus, Neptune, and Pluto represent the higher octave, "more spiritual" counterparts of Mercury, Venus, and Mars respectively. An alternate approach has been to "see the light" and adopt the ancient Hindu model in which there is only one planetary ruler per astrological sign.

In either case, there are strong astrological arguments for the existence of at least two additional planets. The second "octave" of planets lacks two planets of being filled, and physical realization of the Hindu scheme cannot be achieved until the dual rulerships of Mercury and Venus are ended by the discovery of two more planets (see Figure 11.1). Thus, the astrologers have gone boldly forth to search out new worlds. Their results provide us with some of the finest tragicomedy to be found anywhere in literature. Astrologer-psychic Jeanne Dixon, for example, speculates:

"I see a sister planet − hidden by the sun − being discovered by the end of this century. Scientists will land instruments on the planet Jupiter to gain a bird's-eye view of this as yet unknown planet." (11.13).

(Flying saucer bluffs will instantly recognize this as the mysterious "far side of the sun" planet Clarion.) Clearly such a planet cannot exist without producing easily detectable gravitational effects on the motions of the planets Mercury, Venus, and Mars. Moreover, in the two-way tug-of-war that is Newton's Third Law of Motion, the alleged planet Clarion itself would be very quickly perturbed out of a direct earth-sun line, at which time it would shine in our night sky more brightly than any other celestial object, save the sun, moon and Venus. Using an IBM 360/40 computer, Dr. R.L. Duncombe of the U.S. Naval Observatory has in fact demonstrated that Clarion's perturbations on Venus and Mars would reach detectable amounts within a few months and Clarion itself would have been perturbed over 2 degrees away from an earth-sun line in a little over a century, and thus should easily be visible at times of total solar eclipse (11.14).

A second astrological choice for an unknown planet is the so-called planet Lilith, a planet which, in the words of Harold Stone:

" . . . eludes definite knowledge . . . and may one day be known as the ruler of Libra." (11.15).

To track down Lilith's origins, one must go back to 1846, when the French astronomer, Frederic Petit,* the director of the Observatory of Toulouse, appeared before the French Academy and presented a paper (11.16) in which he stated that a second moon of the earth had been discovered by three observers, Lebon and Dassier at Toulouse and Lariviere at Atenac, a town 42 km from

*An astronomer whom Willy Ley describes as "not one of the famous astronomers of his time". (11.17)

161

Toulon, in the early evening hours of March 21, 1846. Petit's calculated orbit for Lilith carried the satellite from an apogee of some 3550 km above the earth's surface to a perigree of 11 km above sea level! Lilith would thus have faded from the memory of astronomers and astrologers alike, were it not for the fact that a young French writer named Jules Verne, found Petit's abstract and in 1865 immortalized Lilith as the vehicle by which he gravitationally deflects his astronaut heros in *Autour de la Lune* from a collision course with the lunar surface. Despite Verne's romanticizing, however, the various theoretical and observational investigations of the possibility of a second natural satellite of the earth which have been made over the past century have all come up emptyhanded (11.18).

Some final words on Lilith, however, are worth mentioning. Parker and Parker in their astrological discussion of the solar system include the following passage on Lilith:

> "Still more doubtful is both the existence and influence of Lilith, said to be a satellite of Earth, one-quarter the size of our Moon, which has been given many mysterious meanings by astrologers in the past. Reputable modern astrologers treat it as skeptically as astronomers: it seems likely that 18th century astrologers may simply have been confused by asteroid 1181 and mistaken it for a more important body." (11.19).

For the record, asteroid 1181 was, according to Pilcher and Meeus (11.20), discovered at Algiers by B. Jekhovsky on February 11, 1927. The object was given a provisional designation of 1927 CQ and named . . . Lilith!

Another planetary visitor to astrological haunts is the hoary old ghost of the intra-Mercurial planet Vulcan. Long ago left for dead by virtually all of the astronomical community, Vulcan is very much alive and well in the astrological literature. Linda Goodman tells us, for instance:

> "It's important to mention here the still unseen planet Vulcan, the true ruler of Virgo, since its discovery is said to be imminent . . . Many astrologers feel that Vulcan, the planet of thunder, will become visible through telescopes in a few years." (11.21).

Manly Hall is even more definitive:

> "An intra-Mercurial planet is not only a possibility but an actual certainty. L.H. Weston, in his brochure, 'The Planet Vulcan,' lists twelve instances where the planet Vulcan has actually been seen by reputable astronomers . . . The position of Vulcan causes its influence to be overshadowed by the power of the sun, and in common with Uranus and Neptune, its effects are regarded as eccentric and malefic." (11.22).

The actual story of the rise and fall of Vulcan is magnificently summarized by Norwood Hanson (11.23) and is worth retelling here if for no other reason than it represents high scientific drama at its finest.

In the decades that followed Newton, the motion of the planet Mercury had long been a bane to mathematical astronomy. Leverrier himself said of the.

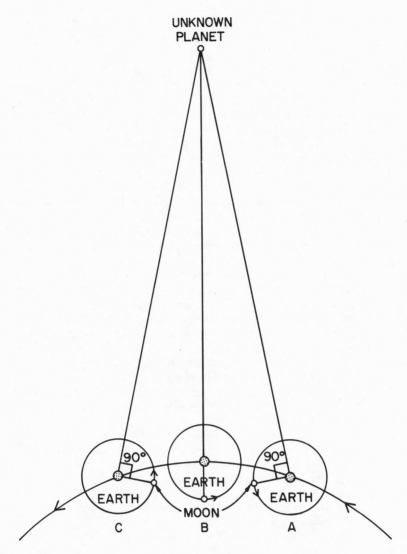

Figure 11.5

Possible use of the "disturbing" square (90°) and opposition (180°) aspects to locate the approximate position of a new planet. The orbital location of the earth and moon as the planet undergoes the three successive "disturbing" aspects with the moon at A (separating square), B (opposition), and C (applying square) unambiguously sets both the planet's line of sight position and its distance from the earth within the orb limits of the aspects.

planet:

"No planet has exacted more pain and trouble of astronomers than Mercury, and has awarded them with so much anxiety and so many obstacles." (11.24).

The basic problem was that like Uranus, the planet Mercury did not move exactly as predicted by Newtonian orbital theory. In essence, the line joining the near and far points of Mercury's orbit about the sun, or so-called line of apsides, slowly changes its orientation in space at a rate of 1.55 degrees per century. Most of this effect could be accounted for in terms of known effects from Newtonian theory, but some 43 arcseconds* or less than 1% of the total could not. The discrepancy is not large, but it is nevertheless real and could not be described by the sum of the forces known to be acting on Mercury. Thus, Leverrier attempted to resolve the discrepancy in exactly the same way in which he had been so triumphantly successful in the case of Uranus. He postulated the existence of a gravitationally perturbing intra-Mercurial planet, and published his predictions in 1859. Almost immediately reports of sightings of the intra-Mercurial planet streamed in from all over the world, virtually all of which were tendered by amateur observers. Most of the observations were quickly dismissed as being due to hoaxes or wishful thinking at the telescope, but the few observations that seemed to be both honestly and skillfully made led Leverrier to accept the existence of the new planet in 1860, and it was appropriately named for Vulcan, the Roman god of fire and metal-working.

The Vulcan "triumph" however, very quickly turned to ashes. The accepted observations of the supposed planet could not be confirmed. Moreover, the constraints that they placed on Vulcan rendered the planet incapable of producing the observed discrepancies in Mercury's motion. By the early 1880's, the matter had been at least observationally closed. There was no planet of any significant size moving inside of the orbit of Mercury. But if the intra-Mercurial planet had quietly exited the scientific world, the perturbations on Mercury's motion that led to its supposed existence in the first place had not. Various other suggestions, such as intra-Mercurial belts of asteroids and rings of dust, suffered the same observational fate as Vulcan. By the turn of the century it had become obvious that Newtonian orbital mechanics simply could not adequately deal with the problem of Mercury's motion. Of course, when such occurs in science, it becomes necessary to develop a new theory or representation that *is* able to adequately deal with such discrepancies. In 1916, Einstein's Theory of General Relativity made the grade and the anomalies in the motion of the planet Mercury could finally be theoretically accounted for. Thus, after withstanding the assaults of the larger and more distant planets in the solar system, the limits of Newtonian mechanics were revealed in the end by one of the tiniest of the planetary worlds. General Relativity was in, Vulcan was out — except in the world of the astrologer.

Perhaps thinking that there is safety in numbers, astrologer Alfred Witte in the 1920's filled out not only the Uranus octave, but the next highest one as well

*The angle subtended by a quarter at a distance of roughly 115 meters.

TABLE 11.1 – A PARTIAL EPHEMERIS FOR TRANS-PLUTO
1975 – 1978*

DATE	ASTROLOGICAL POSITION	CELESTIAL LONGITUDE
Jan. 1, 1975	16° 00' Leo	156° 00'
July 1, 1975	15 00 Leo	155 00
Jan. 1, 1976	16 15 Leo	156 15
July 1, 1976	15 45 Leo	155 45
Jan. 1, 1977	17 50 Leo	157 50
July 1, 1977	16 00 Leo	156 00
Jan. 1, 1978	17 00 Leo	157 00

*American Federation of Astrologers

165

with eight "new" planets, including Cupido, Hades, Zeus, Chronos, Apollo, Admentos, Vulcan and Poseidon (11.25, 11.26). The discovery of Pluto seems to have had the effect of bumping Poseidon into a still higher octave. These planets are still in use today by the so-called Hamburg School of astrologers. While this sytem enjoys considerable support in Austria and Germany, there seems to be very little inclination on the part of astrologers elsewhere to adopt it, particularly since none of these supposed planets to date has been detected by astronomers.

Perhaps the most reasonable claim for a new planet comes to us from the cosmobiologist Reinhold Ebertin who postulated the existence of an object called "Trans-Pluto", an ephemeris for which is currently available from the American Federation of Astrologers (see Table 11.1). Unfortunately, Ebertin's contribution to the planetary worlds of the astrologer lies well inside many of the patrol zones of the various photographic searches which have been conducted for comets, asteroids, variable stars, etc. In being so located, Trans-Pluto, alas, has suffered the same scientific fate as Cupido *et al.*

Despite the withering scientific fire from Schmidt patrol plates, astrologers stubbornly stand by their guns:

"For although such planets as the hypothetical Cupido are almost certainly non-existent, it is just possible — if extremely unlikely — that Witte had identified certain points in the solar system which like the angles that form the aspects in a horoscope, have some unexplained astrological importance." (11.27).

Such commentary, however, cannot hide the fact that for centuries as they trundled around the sun, Uranus, Neptune, and Pluto all offered astrologers a *triumphus astrologius* back into the realm of scientific respectability. When the astrologers failed, Herschel, Adams, Leverrier, and all the others claimed the planets for the side of mathematical and observational astronomy. Even after being stung by three new planets inside of two centuries, the *triumphus* still awaits the astrologers; they need only to successfully predict the existence and location of one or more new planets using *astrological* principles. But instead of presenting us with a Neptune as did Adams and Leverrier, the astrologers have thus far handed us Clarion, Cupido, Lilith, and Vulcan, all of which represent little more than desperate hedges against the ever-present possibility that additional planets may even now be lurking in the heavens waiting for *their* chance to extend short-deck astrology a few more decades.

11.1   Goodavage, J., "Seven by Seven", Signet Books, New York (1978).

11.2   Jobes, G. and Jobes, J., p. 81, "Outer Space: Myths, Name Meanings, Calendars", Scarecrow Press, Inc., New York (1964).

11.3   West, J., and Toonder, J., p. 139, "The Case for Astrology", Penguin Books, Inc., Baltimore, Maryland (1973).

11.4   Alexander, A., p. 26, "The Planet Uranus", American Elsevier Publishing Company, Inc., New York (1965).

11.5   Jones, H., "John Couch Adams and the Discovery of Neptune", MacMillan & Co., New York (1947).

11.6   Grosser, M., "The Discovery of Neptune", Harvard University Press, Cambridge, Massachussetts (1962).

11.7   Shulman, S., p. 115, "The Encyclopedia of Astrology", Hamlyn Publishing Group Ltd., New York (1976).

11.8   MacNeice, L., pp. 182-183, "Astrology", Doubleday and Company, Inc., Garden City, New York (1964).

11.9   Omarr, S., p. 272, "My World of Astrology", Wilshire Book Company, North Hollywood, California (1975).

11.10   Ibid., Ref. 11.3, pp. 140-141.

11.11   Goodman, L., p. 203, "Linda Goodman's Sun-Signs", Bantam Books, Inc., New York (1968).

11.12   Browning, N., p. 190, "Omarr: Astrology and the Man", Signet Books, New York (1977).

11.13   Dixon, J., "Family Weekly", June 23, 1974, p.1.

11.14   Duncombe, R.L., pp. 853-854, Appendix E. "Scientific Study of Unidentified Flying Objects", by E.U. Condon, E.P. Dutton & Co., Inc., New York (1969).

11.15   Stone, H., "Astrology – Your Daily Horoscope", March 1977, p. 40.

11.16   Petit, F., Comptes Rendus, October 12, 1846.

11.17   Ley, W., p. 258, "Watchers of the Skies", The Viking Press, New York (1963).

11.18   Ibid., Ref. 11.14, pp. 258-268.

11.19   Parker, D. and Parker, J., p. 72, "The Compleat Astrologer", McGraw-Hill Book Company, New York (1971).

11.20   Pilcher, F. and Meeus, J., p. 55, "Tables of Minor Planets", Illinois College, Jacksonville, Illinois (1973).

11.21   Ibid, Ref. 11.11, p. 203.

11.22   Hall, M., p. 137, "Astrological Keywords", Littlefield, Adams and Company, Totowa, New Jersey (1975).

11.23   Hanson, N., Isis, 53, 359 (1962).

11.25   Ibid., p. 365.

11.26   King, F., p. 78, "The Cosmic Influence", Doubleday and Company Inc., Garden City, New York (1976).

11.27   Tyl, N., pp. 133-167, "Times to Come", Vol. XII, "The Principles and Practice of Astrology", Llewellyn Publications, St. Paul, Minnesota (1974).

11.28   Ibid., Ref. 11.25, p. 80.

Chapter 12

The Astrological Catastrophe

The "bottom line" in any scientific description of a phenomenon in nature is its ability to predict future events or new experimental results. The scientific method, in fact, demands that such descriptions are constantly probed, prodded, and poked with a predictive stick if any scientific progress is to ensue. The ultimate success of a given theory is judged directly in accordance with its ability to render successful predictions. Descriptions which are not able to predict accurately find themselves in the dustbin of scientific history. Such was the fate, for example, of William Herschel's description of sunspots as openings leading into a cool and inhabited solar interior. On the other hand, descriptions which can predict correctly are crowned with the rewards that success brings. Thus the seemingly Alice-in-Wonderland qualities of quantum mechanics were regarded with a great deal of scientific suspicion early in this century, but they ultimately carried the day because of their predictive success. It is this aspect of the scientific method that provide the astrologer with one last opportunity to demonstrate the veracity of their methodology, namely through the accuracy of the predictions rendered by that methodolgy. In other words, given a free hand to cast and interpret an entire horoscope by whatever means they choose, how do astrologers fare when asked to make specific predictions based on their experience and knowledge of the astrological correspondences?

Unfortunately, attempts to answer this highly crucial question are almost always made in terms of a few examples, usually preselected in favor of the given point of view. Thus, "world famous" astrologer Eva Petulengro is billed as the successful astrologer:

> "who accurately predicted Senator Ted Kennedy's announcement that he wouldn't run for president and that Richard Nixon would resign without being impeached . . . " (12.1),

168

while scientists respond with a selected list of their own favorite predictive misfires by astrologers. Such as:

Sybil Leek's predictive choices for the 1976 presidential candidates were Ronald Reagan and Edward Kennedy (12.2).

The aforementioned Eva Petulengro predicted that by the end of 1975, "North Korea — backed by Red China — will take over South Korea without any interference from the U.S." (12.1).

Irene Hughes predicted an attempt on President Carter's life on the very day of his inauguration (12.3).

Carroll Righter predicted that in February of 1978 President Carter will "be the victim of the most daring physical attack ever made on a president". Mr. Righter also saw "something such as explosion in the White House or an attempt on his life with a firearm" (12.4).

Such argumentation, whether it comes from the scientific or astrological side of the aisle, offers no real solution to the question. In an attempt to minimize such "proof by example" kinds of arguments, we have, for a number of years, kept track of specific astrological predictions which have been made in the printed media. In each case the prediction was rated as unsuccessful if the event did not occur within the time limit prescribed by the astrologers themselves. A total of 3011 predictions were recorded and of those, 338 or 11 percent came through as advertised (Table 12.1). In each case, any prediction which could have been attributed to shrewd guesses (the SALT talks will continue to be stalled for another year), vague wording (there will be a tragedy in the eastern United States during the spring months), or "inside" information regarding the person(s) involved (starlet A will be married to director B before Christmas) were all counted in the astrologers' "successful" columns. No astrologer was listed individually unless we could find 100 or more of their predictions, and no astrologers' results were included in the "Miscellaneous Astrologers" category unless he or she tendered ten or more predictions for the public media. Under these conditions, no astrologer included in this survey was able to correctly predict more than 20 percent of the time, and most experienced considerably smaller rates of success. It should be recognized, of course, that such surveys span a number of different branches of astrology, including, in particular, natal, mundane, and horary astrology. Nevertheless, we would hope that those astrologers who profered their predictions for public consumption would not have done so unless they themselves felt competent to do so in the particular area of astrology that is involved.

The results in Table 12.1 paint a dismal picture indeed for the traditional astrological claim that "astrology works". Of the over three thousand predictions made by astrologers, who in casting the required horoscopes presumably took all of the various chart factors and weights into proper astrological account, barely 10 percent were fulfilled. Moreover, included in this ten percent are the aforementioned "marginally correct" predictions in which the astrologer was given the benefit of the doubt, thus making the observed 11 percent predictive accuracy more of an upper limit value rather than an average

169

TABLE 12.1   Predictive Success of Leading Astrologers
And Astrological Publications

Predictor	Total Predictions Recorded	Predictions Fulfilled	Percent Correct
American Astrology Magazine*	462	71	15
Astrology Magazine	544	62	11
Jeanne Dixon	134	14	10
Horoscope Magazine*	509	54	11
Horoscope Guide*	532	68	13
Sybil Leek	128	6	5
Sydney Omarr's Astrological Guide**	118	13	11
Carroll Righter	123	11	9
Miscellaneous Astrologers‡	461	39	8
TOTALS	3011	338	11

*Time period from January 1974 – March 1979
**1974, 1975, 1976, 1977, 1978 editions
‡Minimum of ten predictions/astrologer

170

value. At this point, the stark contrast between the accomplishments of modern astronomy versus those of modern astrology is perhaps best illustrated by a simple comparison of the results in Table 12.1 with the 100 percent success rate for the many thousands of predictions annually made by astronomers for the positions, configurations, and behavior of celestial objects (See Figure 12.1).

This is not to say that scientists have not had their share of predictive troubles. Perhaps the classic example in this regard came at the end of the last century. At that time physicists were interested in explaining the behavior of radiant energy coming from a glowing or incandescent source of light. If the light from such a source is carefully measured, one finds that the emitted energy has a non-uniform distribution by wavelength or color. Thus the total energy of the red wavelengths of the emitted light will in general not be the same as that of the blue or yellow wavelengths. Observations of the energy emitted at each of the known wavelengths of the electromagnetic spectrum reveals that the final distribution by wavelength takes on the hump-shape shown by the solid line in Figure 12.2. The actual location of the peak energy and the height of the curve above the wavelength axis will change with the object's temperature but the energy is found to approach zero at both very large and very small wavelengths regardless of the temperature. This so-called black-body* curve is thus an experimentally observed fact, just as an event in national or international affairs would be an observed fact for the astrologer. Using the theoretical descriptions of radiant energy then available, physicists, however, had predicted beforehand that the amount of energy emitted at successively smaller wavelengths should correspondingly increase to infinitely large values as indicated by the dashed line in Figure 12.2. In short, the physicists had predicted that the black-body should rise toward infinity at very small wavelengths, while the observed experimental results indicated that it dips back toward zero. It is difficult to imagine a discrepancy larger than that between zero and infinity! The scientists did not totally lose their sense of humor over all of this; the effect came to be known as the "ultraviolet catastrophe".

The response to the "ultraviolet catastrophe" required by the scientific method was, of course, the development of a set of laws (Planck's Law, Wien's Law, etc.) and a theory (quantum theory) which would bring the theoretical predictions regarding the behavior of energy radiating from an incandescent source in line with experimental results.

There is certainly nothing new about such discrepancies between theoretical prediction and empirical observation in science. They always have been present, are currently present, and, in all probability, will continue to be present well into the foreseeable future. Indeed, their presence and the attempts to scientifically resolve them, as in the case, for example, of the current debate over red-shifted spectra of galaxies (12.5), generate the finest moments in scientific advance.

*In physics and astronomy a black-body is an idealized object which lacks the ability to reflect electromagnetic energy. A black-body is thus a perfect absorber and radiator of electromagnetic energy.

171

PHENOMENA, 1977

CONFIGURATIONS OF SUN, MOON AND PLANETS

d h		
Jan. 1 02	Jupiter 0°·8 N. of Moon	Occn.
3 10	Earth at perihelion	
5 12	FULL MOON	
6 08	Mercury in inferior conjunction	
8 00	Saturn 6° N. of Moon	
9 02	Vesta stationary	
12 11	Mercury 4° N. of Mars	
12 20	LAST QUARTER	
14 04	Uranus 0°·6 S. of Moon	Occn.
15 20	Jupiter stationary	
16 10	Moon at perigee	
16 12	Neptune 2° S. of Moon	
17 07	Mercury stationary	
18 01	Mercury 2° S. of Moon	
18 11	Mars 6° S. of Moon	
19 14	NEW MOON	
23 09	Pluto stationary	
23 11	Venus 3° S. of Moon	
24 11	Venus greatest elong. E. (47°)	
27 05	FIRST QUARTER	
28 06	Moon at apogee	
28 10	Jupiter 1° N. of Moon	
29 00	Mercury greatest elong. W. (25°)	
Feb. 2 10	Saturn at opposition	
4 04	FULL MOON	
4 04	Saturn 6° N. of Moon	
10 10	Uranus 0°·9 S. of Moon	Occn.
10 12	Pallas at opposition	
11 04	LAST QUARTER	
11 04	Moon at perigee	
11 19	Ceres stationary	
12 19	Mercury 0°·1 S. of Mars	
12 20	Neptune 0°·1 S. of Moon	
14 22	Uranus stationary	
16 12	Mars 6° S. of Moon	
16 17	Mercury 7° S. of Moon	
18 04	NEW MOON	
21 17	Venus 3° N. of Moon	
24 22	Jupiter 2° N. of Moon	
25 03	Moon at apogee	
26 03	FIRST QUARTER	
26 09	Vesta stationary	
Mar. 1 09	Saturn 6° N. of Moon	
3 09		
5 17	FULL MOON	
8 33	Moon at perigee	
9 15	Uranus 1° S. of Moon	Occn.
11 07	Pallas stationary	

d h		
Mar. 12 02	Neptune 3° S. of Moon	
12 12	LAST QUARTER	
14 19	Venus stationary	
16 05	Mercury in superior conjunction	
17 11	Mars 6° S. of Moon	
18 11	Neptune stationary	
19 19	NEW MOON	
20 18	Equinox	
21 07	Juno stationary	
21 13	Venus 8° N. of Moon	
24 15	Jupiter 2° N. of Moon	
24 20	Ceres at opposition	
24 22	Moon at apogee	
27 19	Mercury 8° S. of Venus	
27 22	FIRST QUARTER	
30 17	Saturn 6° N. of Moon	
Apr. 2 16	Pluto at opposition	
4 04	FULL MOON	Eclipse
5 21	Moon at perigee	
5 22	Mars 4° S. of Moon	
6 06	Venus in inferior conjunction	
8 08	Neptune 3° S. of Moon	
10 16	Mercury greatest elong. E. (19°)	
10 10	LAST QUARTER	
11 12	Saturn stationary	
15 12	Mars 4° S. of Moon	
16 20	Venus 5° N. of Moon	
18 11	NEW MOON	
19 16	Mercury 5° N. of Moon	
20 10	Mercury stationary	
21 09	Jupiter 3° N. of Moon	
21 12	Moon at apogee	
24 21	Venus stationary	
26 15	FIRST QUARTER	
27 01	Saturn 6° N. of Moon	
30 06	Uranus at opposition	
30 17	Mercury in inferior conjunction	
May 3 07	Uranus 1° S. of Moon	Occn.
3 13	FULL MOON	Eclipse
4 05	Moon at perigee	
5 16	Neptune 3° S. of Moon	
10 04	LAST QUARTER	
11 23	Venus greatest brilliancy	
13 00	Mercury stationary	
13 18	Venus 1°·3 of Mars	
14 11	Venus 1°·3 of Moon	
14 14	Mars 2° S. of Moon	

PHENOMENA, 1977

CONFIGURATIONS OF SUN, MOON AND PLANETS

d h		
May 16 07	Mercury 2° S. of Moon	
17 06	Ceres stationary	
18 03	NEW MOON	
18 18	Moon at apogee	
20 13	Jupiter 3° N. of Moon	
24 11	Saturn 6° N. of Moon	
26 03	FIRST QUARTER	
27 23	Mercury greatest elong. W. (25°)	
30 16	Uranus 0°·9 S. of Moon	Occn.
June 1 15	Moon at perigee	
1 21	FULL MOON	
2 02	Neptune 2° S. of Moon	
3 13	Venus 1°·2 S. of Mars	
4 10	Jupiter in conjunction with Sun	
8 15	Neptune at opposition	
8 15	LAST QUARTER	
12 11	Mars 0°·1 N. of Moon	
12 15	Venus 2° S. of Moon	
14 21	Moon at apogee	
15 07	Venus greatest elong. W. (46°)	
15 07	Mercury 5° N. of Aldebaran	
16 18	NEW MOON	
20 07	Mercury 0°·1 N. of Jupiter	
20 21	Saturn 6° N. of Moon	
21 12	Solstice	
24 13	FIRST QUARTER	
27 00	Uranus 1° S. of Moon	Occn.
28 21	Pluto stationary	
29 11	Neptune 2° S. of Moon	
30 00	Moon at perigee	
30 00	Mercury in superior conjunction	
July 1 03	FULL MOON	
5 20	Earth at aphelion	
8 05	LAST QUARTER	
11 11	Mars 2° N. of Moon	
12 08	Moon at apogee	
12 09	Venus 1° N. of Moon	
12 20	Juno stationary	
13 19	Jupiter 4° N. of Moon	
15 19	Venus 3° N. of Aldebaran	
16 14	NEW MOON	
16 14	Uranus stationary	
18 03	Saturn 6° N. of Moon	
20 01	Mercury 0°·4 N. of Saturn	
23 20	FIRST QUARTER	

d h		
July 24 07	Uranus 1° S. of Moon	Occn.
26 19	Neptune 3° S. of Moon	
28 02	Moon at perigee	
28 03	Mercury 0°·1 S. of Regulus	
30 06	Venus 1°·6 S. of Jupiter	
30 11	FULL MOON	
Aug. 1 12	Mars 5° N. of Aldebaran	
6 21	LAST QUARTER	
8 20	Moon at apogee	
9 00	Saturn 6° N. of Moon	
9 11	Mars 4° N. of Moon	
10 13	Jupiter 4° N. of Moon	
11 14	Venus 4° N. of Moon	
13 06	Saturn in conjunction with Sun	
14 22	NEW MOON	
16 23	Mercury 0°·9 S. of Moon	Occn.
20 13	Uranus 2° S. of Moon	
21 23	Mercury stationary	
22 01	FIRST QUARTER	
23 01	Neptune 3° S. of Moon	
23 17	Venus 7° S. of Pollux	
24 09	Moon at perigee	
25 17	Neptune stationary	
28 20	FULL MOON	
Sept. 4 22	Mars 0°·5 S. of Jupiter	
5 06	Mercury in inferior conjunction	
5 15	LAST QUARTER	
5 18	Moon at apogee	
7 07	Jupiter 5° N. of Moon	
7 09	Mars 5° N. of Moon	
10 21	Venus 5° N. of Moon	
11 13	Saturn 5° N. of Moon	
13 09	NEW MOON	
13 19	Mercury stationary	
15 08	Vesta in conjunction with Sun	
16 21	Uranus 2° S. of Moon	
18 00	Moon at perigee	
18 13	Venus 0°·4 S. of Saturn	
19 07	Neptune 3° S. of Moon	
20 06	FIRST QUARTER	
21 08	Mercury greatest elong. W. (18°)	
22 03	Venus 0°·4 N. of Regulus	
23 04	Equinox	
27 08	FULL MOON	Penumbral Eclipse
Oct. 3 14	Moon at apogee	
4 21	Jupiter 5° N. of Moon	
5 01	LAST QUARTER	

Figure 12.1 Two hundred astronomical predictions from the pages of the "American Ephemeris and Nautical Almanac"

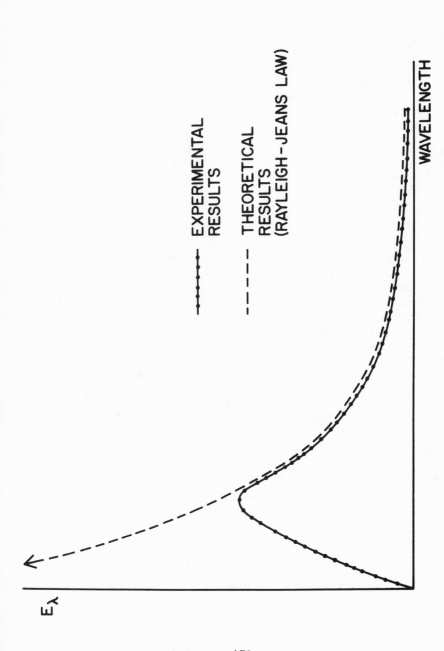

Figure 12.2   A schematic diagram of the Ultraviolet Catastrophe. The wide disagreement between the radiative behavior of a black body predicted by the Rayleigh-Jeans Law (dashed line) and the actual observed behavior (connected dots) helped usher in modern quantum theory in physics.

173

But how do astrologers respond when confronted with predictive "astrological catastrophes" of their own? A piece done a few years ago by Thomas Buckley (12.6) for the "New York Times Magazine" provides us with some preliminary insights into the matter. It seems that in 1968 a number of leading astrologers in the U.S. predicted that the then Jacqueline Kennedy would not marry in 1968. Upon her marriage to Aristotle Onassis on October 20, 1968, the responses of the astrologers involved included the following:

> Jeanne Dixon ordered her syndicated astrological column, scheduled for October 20, 1968, withdrawn from publication and replaced it with a corrected second version. In that first column she had written, "I still stand on my New Year's predictions and see no marriage for Jackie Kennedy in the near future".

> Sydney Omarr claimed incredibly that he was at least partially right in his prediction because, "I predicted a *divorce* (italics ours) would make international headlines about that time and that both marriage and divorce are in the seventh house.

> Most interesting of all was Zoltan Mason's response: "As far as I'm concerned, there is no marriage. A 69-year old man! That is not a marriage".

This is hardly the stuff of which searches for truth are made. In fact, as we journey through the marvelously psychedelic world of astrological prediction we find that the most clever, intelligent, and successful of the astrologers have adopted instead a variety of ploys designed specifically for reducing or eliminating the impact of their missed predictions.

Perhaps the oldest and most traditional among such methods is to clothe the predictions in the most confusing, subjective, and nebulous terms possible, thus allowing for a wide variety of interpretations. Astrologer Ronald Davison best lays it out for us in his advice to budding astrologers when he says "Never attempt to delineate or predict in too much detail" (12.7). And, heeding Mr. Davison's advice, astrologers as often as not do indeed provide us with delineations devoid of detail.

In checking the "Tomorrow's News" section of "American Astrology" magazine prior to the dramatic and historic visit of President Anwar Sadat to Israel in November of 1977, for example, we find under Israel the following prediction for November of 1977:

> "...Knesset activity will dominate the month. There *can** be a *promise* of peace or peace negotiations ... shows the government to be in high favor and popular, but ... one member of the Cabinet *could* leave ... *may* bring the death of soldiers and possibly a general ... bringing the nation into world prominence ... Peace talks *can* be suspended and any negotiations are held secretly. Knesset now dithers and divides. A new militancy is popular and financial stringency will be enforced. Sweeping fervor for the land and country is upheld by

*Italics ours

174

workers and all the services, civil and military. A determined effort at diplomacy is made by people from other countries, but the time and mood has changed." (12.8)

It would seem that an occasion of the order of magnitude of Mr. Sadat's visit to Israel (he was named "Time" magazine's Man of the Year) should have elicited a much more enthusiastic and detailed prediction than the editors of "American Astrology" saw fit to provide us with.

Nor did the other astrological publications fare any better. Robert Shannon, for example, wrote a five page article for the May 1978 issue of "Astrology — Your Daily Horoscope" which was entitled, "Israel: Portrait of a Nation Reborn". Not one reference was made at any point to a possible Sadat initiative, despite the fact that Mr. Shannon identifies several periods of potential crisis as well as goodwill for Israel (12.9). Ironically the editors of the same magazine inserted a two page "update" in the same issue in which the astrological forces behind the Sadat initiative were set forth in considerable detail — after the fact (12.9).

There can also be a vagueness in the time coordinate of a given prediction. In 1977, for example, Carroll Righter predicted:

"A disclosure about UFOs will startle the world. The President will release detailed government records which will prove conclusively that we have been visited by beings from other planets. This will happen after the appearance of an unprecedented number of UFOs over major cities." (12.10)

When may we expect these spectacular events? Mr. Righter never tells us.

Closely related is the technique of predictions of events which are to occur so far into the future that it will either be forgotten or nobody will care when the time is at hand for their fulfillment. Thus Nostradamus (12.11) predicted the end of the world* would be in 1999, and more recently, world-wide catastrophes are currently scheduled for 1982 (12.12), 1984 (12.13), and 2011 (12.14). The danger with such a technique is, of course, that someone *will* remember and/or care. Astrologer Katina Theodossiou, for example, predicted in 1959 that:

"The next period of peak prosperity in the U.S. will be in the mid 1970's". (12.15)

Thus the old "only time will tell" refrain appears to be the only empirical response that can be mounted in the face of such techniques.

A second predictive technique often employed by astrologers is the use of simple common sense. When called upon in the summer of 1976 to make a prediction on the inaugural address of January 1977, astrologer Doris Thompson responded:

"I predict that the inaugural address will contain reference to: ecomomics, foreign policy, oil, law and order, equal rights, hard work,

---

*Or, at the very least a universal upheaval, depending on the translator.

175

TABLE 12.2   The Most Common Astrological Predictions From the Table 12.1 Survey*

Prediction	Number of Times Appearing
Assault on the President	26
Contact with Alien Beings	23
Medical Breakthrough in Cancer Research	22
Marriage of Jacqueline Onassis	19
Divorce of Farrah Fawcett-Majors	17
War in the Middle East	14
Medical Breakthrough in Heart Disease	11
Economic Hard Times	11
Economic Good Times	10

*Only those predictions appearing ten or more times are listed.

176

group efforts, technology, and THERE IS THE POSSIBILITY THAT HE MAY INTRODUCE A BLACK WOMAN TO THE UNITED STATES AS A MEMBER OF HIS CABINET *". (12.16)

Clearly astrologer Thompson is in good shape with her prediction, not because of any astrological correspondences, but simply because the issues mentioned from economics to participation in the government by women and minorities are among the most important issues of the day for the United States. We would expect from common sense alone, therefore, that an incoming President would indeed at least make reference to them in an address of this importance.

The most visible of the astrologers such as Carroll Righter and Sydney Omarr also enjoy a certain built-in predictive advantage which arises from their close association with well-known celebrities. Thus such an astrologer might be privy to inside information regarding various celebrities' careers, marriages, etc. Clearly such predictions could just as easily be made by a skilled reporter or gossip columnist operating without the benefit of sidereal sources.

One of the most interesting and unexpected results that came from the analysis summarized in Table 12.1 is the rather large number of repeated predictions. Apparently the tack taken here is to continue time after time to predict a single occurence such as an assassination, medical breakthrough, etc. with the hope that ultimately it will come true (see Table 12.2). It is a tactic that is particularly insulting to the general public, since it assumes that people will tend to forget the number of times the given prediction was made and then went awry and remember only the one time that the given prediction "hit".

We have also found that astrologers quite often preface their predictions with enough disclaimers to make an insurance salesman blush. The last paragraph to the "Commodity Futures Trading: 1976" section of "Sydney Omarr's Astrological Guide for You in 1976" reads in part:

> "All rules have exceptions, of course . . . don't risk money you can't afford to lose. 'Scared money doesn't make money' . . . don't think of either stocks or commodities as gambling . . . " (12.17)

Such statements are not surprising in view of the fact that in the course of comparing the predictions contained in that section with data contained in the 1976 and 1977 Commodity Yearbooks (12.18), we found that co-authors Omarr and Richardson could have done just as well by basing their "up" and "down" commodities predictions on the toss of a coin.

In recent years a more interesting tack employed by astrologers is to simply get out of the prediction business altogether. The idea here is that prediction now has "no place" in modern astrology, since, in the words of West and Toonder:

> " . . . nor would astrologers like to take bets on the statistical accuracy of their predictions. But the one strong point of modern astrology, the one aspect of it upon which astrologers are consistently willing to stake their reputations in public, is their ability to analyze character upon the

*Caps are Ms. Thompson's

177

basis of the horoscope." (12.19)

In other words, the role of the astrologer is now seen as one of a sort of sidereal psychiatrist, psychologist, or vocational guidance counselor whose insights in these areas are far superior to their colleagues in these same areas who have not seen the astrological light. But how true is this claim? Unfortunately there is currently no overwhelming evidence in favor of either side. In the 1950's, for example, the American psychologist Vernon Clark tested the abilities of 23 astrologers to match individual horoscopes and various qualities of the respective individuals such as career, etc. In three separate tests of this nature, the astrologers claimed to successfully match person and horoscope at or above the .01 level of statistical significance (12.20). Later attempts to duplicate Clark's results, however, have resulted in considerably lower levels of statistical significance, with the X^2 p values obtained ranging from .004 to 0.38 (12.21). Gauquelin (12.22), in support of these latter results, found that not one astrologer he tested could statistically pass the so-called "test of opposed destinies", in which the astrologer is called upon to horoscopically separate twenty well-known criminals from twenty people who led long and peaceful lives.

Attempts to test the abilities of astrologers to pick out personality traits have yielded similarly inconclusive results. West and Toonder summarize the astrological explanation for the predictive difficulties:

"Most professional astrologers prefer that clients consult them in person. Astrologizing "blind" as in Clark's experiment is as difficult as diagnosing a disease with the patient absent." (12.23)

This well-known preference on the part of astrologers for dealing with their clients in person, may have a more mundane purpose. On the spot personality analyses are after all really not so difficult with a little practice. Furthermore, in examining the degree of acceptance of horoscope personality descriptions, psychologist C. R. Snyder (12.24) found that to a large extent it depended on situation factors alone rather than any actual relationship between astrological interpretations and an individuals observed personality. Different persons given the identical general horoscope description rated its validity higher the greater their belief that it was derived specifically on the basis of his or her birth datum. If such findings hold up under additional testing, even these most recent claims made for "humanistic" astrology are in for some big trouble.

The method of last resort adopted by astrologers for dealing with the astrological catastrophe, however, is to employ the astrological refrain for all ages, "the stars incline, they do not compel". In other words, after all that is said and done in terms of casting and interpreting the horoscope and the correspondences that it presumably signifies, the astrologer still has a way out of the collapsing building; the heavenly influences can be overriden by one's own free will. The concept represents the final line of predictive retreat for the astrologer, but in a certain sense, it is the most marvelous ploy of all. Suppose, for example, that an astrologer warns a client of an impending automobile accident. As a result of the astrologer's suggestion, the client might have the accident when it otherwise wouldn't have happened, or the accident might happen just in the course of events. If the accident occurs, the astrologer's prediction comes through as proclaimed. If the accident does not occur, however, the astrologer can still credit for placing the client on the alert to the accident's probability, thereby permitting the client to take the necessary steps to avoid it. In either case, the astrologer wins. The concept is particularly bizarre in light of the scientific fact that any physical law developed from experimental results holds up at all times, not just when we deem it convenient. If a mountain climber slips off a cliff, for example, it will take a most persuasive astrological argument indeed to convince the ill-fated climber that the law of gravity inclines but does not compel!

The one response to the astrological catastrophe that is seldom if ever advanced by astrologers is, of course, the complete overhaul of the correspondences and methodology. In light of a 90 percent predictive failure rate, the same scientific method which overturned Newtonian mechanics for a mere 43 arcsecond/century advance in the perihelion point of the planet Mercury, in fact cries out for just such an overhaul. Rather than answering those cries, however, the astrological community has instead developed an impressive array of methods and techniques for the sole purpose of excusing or minimizing their predictive failures without even remotely threatening the basic "truths" of the astrological gestalt. Such is not the road to the scientific credibility and respectability that so many astrologers for so many years have complained about not having.

12.1   *National Enquirer*, July 8, 1975, p. 20.

12.2   Leek, S., "Is the White House in Their Stars?", *Ladies Home Journal*, June, 1976, p. 28.

12.3   Hughes, I., *National Enquirer*, November 23, 1976, p. 16.

12.4   Righter, C., *National Enquirer*, July 26, 1977, p. 33.

12.5   Field, G., Arp. H., and Bahcall, J., "The Redshift Controversy", W.A. Benjamin, Inc., London (1973).

12.6   Buckley, T., *New York Times Magazine*, December 15, 1968, p. 146.

12.7   Davison, R., p. 169, "Astrology", ARCO Publishing Company, Inc., New York (1975).

12.8   *American Astrology*, November, 1977, p. 42.

12.9   Shannon, R., *Astrology: Your Daily Horoscope*, May 1978, p. 48.

12.10   Ibid., Ref. 12.4.

12.11   Russell, E., "Astrology and Prediction", p. 164, B.T. Batsford, Ltd., London (1972).

12.12   Gribbin, J., and Plagemann, S., "The Jupiter Effect", Vintage Books, New York (1974).

12.13 Omarr, S., "Sydney Omarr's Astrological Guide for You in 1977", p. 2, Signet Books, New York (1976).

12.14 Waters, F., "Mexico Mystique", Swallow Press, Inc., Chicago (1975).

12.15 *Newsweek*, March 30, 1959, p. 85.

12.16 Thompson, D., *CAO Times*, Summer Solstice, 1976, p. 46.

12.17 Omarr, S., "Sydney Omarr's Astrological Guide for You in 1976", p. 275, Signet Books, New York (1975).

12.18 "Commodities Yearbook", Commodity Research Bureau, Inc., New York (1976 and 1977 eds.).

12.19 West, J., and Toonder, J., "The Case for Astrology", p. 225, Penguin Books, Inc., Baltimore, Maryland (1973).

12.20 Clark, V., "In Search", Winter 1959/60.

12.21 Dean, G., and Mather, A., "Recent Advances in Natal Astrology", p. 549, The Astrological Association, Bromley Kent, England (1977).

12.22 Gauquelin, M., "The Cosmic Clocks", p. 85, Henry Regnery Company, Chicago (1967).

12.23 Ibid., Ref. 12.18, p. 297.

12.24 Snyder, C., "Why Horoscopes are True: The Effects of Specificity on Acceptance of Astrological Interpretations", J. of Clinical Psychology **30**, 577 (1974).

Chapter 13

The Fundamental Astrological Principle

Throughout the above discussion we have found time and again that traditional astrology has failed to carefully probe what we will refer to as the fundamental astrological principle — the idea that human beings and their actions are influenced at least in part by celestial objects. We have seen that it is difficult to make any sense of the correlations claimed by astrology either from the point of view of testable physical mechanisms or simply in the cyclic patterns revealed through the regular movements of the planets and their occasional close encounters. In any case there is no agreement between astrologers as to which view is correct. Certainly the persistence by the "synchronists" in rejecting the aid of causal connections has created a major obstacle to the acceptance of astrology. But if traditional astrology and astrologers have failed in this endeavor, to what extent, if any, have the orthodox scientific methods succeeded in establishing the existence of "celestial" influences in our lives?

As we have discussed in Chapter 8, science has in fact come to some understanding of a number of external agents that affect this planet and its creatures in a variety of ways. Sunlight and ocean tides, i.e. electromagnetic radiation and gravitational forces, are so obviously a part of our daily lives we seldom give a passing thought to them as arising from extraterrestrial objects. Other effects may be far more subtle, for example, the collision of high energy particles with the earth produces the aurora, and although these particles ordinarily are difficult to detect, they are capable of producing noticeable results such as genetic mutations. It should be kept clearly in mind that these phenomena are very definitely not astrological, even though some astrologers have attempted to argue that they validate traditional astrological precepts. They are as natural as the sunlight falling on grassy fields. They do not by analogy suggest that when Jupiter is suitably placed in your horoscope you will tend to

181

be overweight.

There are just two bodies of immediate interest to us here. The sun, as noted earlier, is an ordinary, run-of-the-mill, common sort of star. It is 300,000 times more massive than the earth, but it is neither especially large or small, nor especially hot or cold. There are literally millions of other stars just like it in the Milky Way. Smaller but closer, the moon is our nearest celestial companion of note, only 383,000 km distant, and so it is a likely suspect as a celestial agent. At moments it may appear ethereal and beautiful; under close scrutiny, it does not seem particularly mysterious. We have afterall been there and we have scratched its surface to find that it is composed of material not so very different from the earth itself.

Other than the reflected sunlight we see on moonlit nights, the next most apparent consequence of having this small rocky neighbor is the existence of tides. We have already shown in Chapter 8 that the tides will not serve any traditional astrological role, but the effects on the earth are very clear. Resulting from the greater gravitational pull on the side closest to the object, tides are produced by both the moon and the sun, the sun being about half as important. Notably, these tides are not just restricted to the familiar effect on the oceans. The solid earth also responds, but with a tidal bulge of its crust that does not get larger than about 40 cm (13.1). Furthermore there are tides in the earth's atmosphere although the effect is slight due to the small mass of the atmosphere. It is scarcely measureable in high and middle latitudes because any atmospheric tide is masked by the far larger changes occurring in weather systems and by temperature variations. It can be detected in the equatorial regions as a small cyclical barometric pressure change of about 2 mm of mercury (normal is 760), a change of about 0.2%. Interestingly this variation follows the *solar* daily period, not the lunar tidal period owing to properties of the earth's atmosphere. The lunar effect is only about 1/20 as large in this case (13.2). One might imagine that there could be further consequences of these tidal forces.

Earthquakes are not as uncommon on the earth as we would like them to be. Each year there are between 18,000 and 22,000 shallow (less than 60 km deep) earthquakes of Richter scale magnitude 2.5 or greater. That's about 55 earthquakes each day over the earth as a whole. They are part of our inheritance from the way the earth formed and its current faulty construction. Most of these quakes occur when the stresses in the earth's crust become greater than the frictional forces which ordinarily prevent different pieces of the crust from sliding against one another. Since there is a "celestial agent" stretching and pulling the earth's crust through the tidal force, it wouldn't be too surprising to find some evidence that the moon is responsible for triggering some earthquakes. There is indeed some evidence of such an effect, but not much. Most attempts to search for such effects have failed (13.3). Part of the reason is that the stress produced by any tidal pull is from 1/100 to 1/1000 the stress normally present in crustal faults, so the tides aren't very effective. The few cases where tidal triggering appears important are those where the pull happens to be just in the right direction to enhance the natural tendency toward slippage (13.3, 13.4).

182

Similarly there is a limited amount of evidence that the time of some volcanic eruptions may be influenced by tidal forces (13.5, 13.6). However, it does not seem to be the semi-daily tidal period because the only correlation is a barely significant one over a two-week period, that is at the first and third quarter lunar phase when the sun and moon pull at right angles to one another. Why this should be so is not presently clear.

In a number of ancient cultures folk-myths have frequently connected the moon with rainfall. The Romans believed there was a relationship between the phases of the moon and rain and humidity. They may have been right. There has been some evidence reported which does suggest a correlation of weather patterns with the lunar cycle. In one study data from more than 1,500 weather stations with records going back as early as 1909 were gathered by D. A. Bradley, W. A. Woodbury, and G. W. Brier (13.7). They found a tendency for days of maximum precipitation to occur near the middle of the first and third weeks of the lunar month especially on the third to fifth day after a new or full moon. Similar results were published at the same time for the southern hemisphere by two Australian meterologists, E. E. Adderly and E. G. Bowen (13.8). Still a third study of the frequency of occurence of thunderstorms found that the peak activity seemed to be about two days after the full moon (13.9). A reason suggested to account for such effects was that the moon might be responsible for changes in the amount of meteoric dust in the atmosphere whose particles can serve as nuclei for the formation of ice crystals and in turn rain droplets.

The raising and lowering of water tides has resulted in the development of complicated behavior patterns in tidal marine life that have yet to be fully understood by biologists. In some organisms this seems to be in fact a sensitivity to the tidal gravitational forces. Oysters and fiddler crabs displaced inland from their normal coastal habitats seem to adjust their activities after a few days so that they are in synchrony with the local tidal forces (13.10). How could this happen when the animals were maintained in dark sealed containers in no way resembling their normal tidal environment? No occult forces need to be invoked. A wide variety of animals have been found to possess gravity sensing devices called statocysts. (13.11). These organs consist of a tiny mass anchored by sensory hairs within the cyst cavity. Small changes in gravitational forces move the tiny mass, and are thus detected by the animal.

Other effects found in apparent luni-solar synchrony range from varying rates of activity in the West Indian scarab bettle (13.12) and the breeding habits of the California grunion (13.13) to varying rates of intestinal calcium transport in the frog (13.14). Peaks or lulls in creative activity are often found around the new or full phase of the moon, and in some cases are likely to be related to the amount of light available at night. A number of experiments with specially-designed bug traps have shown there are as few as one fifth as many insects about at the time of the full moon (13.15). There is a clear survival benefit to be gained by many creatures from restricted activity around full moon when they might be more easily detected by nocturnal predators. There is certainly nothing unnatural here. Indeed, these behavior patterns may be

183

considered just a variant of the normal light/dark cycle of circadian rhythm. It makes sense that a usually nocturnal species might be less active on brightly moonlit nights.

Astrology, however, makes no claims regarding the personalities of moths, so we must turn our attention to other areas. In dealing with human behavior, the situation is considerably more complex since the range in types of activity is so much wider, subtle, and sometimes bizarre. Separating myth from fact is made more difficult by the emotions and attitudes of the observer who is usually a believer or sceptic beforehand — interest so frequently means bias — and the neutral, objective investigator cannot but be rare. It is little wonder then that many of the studies connecting celestial and human factors have yielded inconclusive or contradictory results.

The folklore built around the moon and its effect on people is substantial. One of the more extreme examples is lycanthropy, that is, the werewolf legend, a subject often celebrated on the silver screen from the earliest wolfman to his contemporary descendents. Here the effect of the full moon is quite dramatic to say the very least. We have not seen any references to this sort of thing in the astrological literature, but do not doubt that it has been considered somewhere. To see if there is anything to any of the myths, we shall concentrate our attention on more common phenomena and the few serious studies that have been done using quantitative information.

One of the few investigations apparently supporting a correlation between lunar phase and deviant human behavior was published by Arnold Lieber and Carolyn Sherin in 1972 (13.16). As an indicator of a possible relationship, they choose to examine the times at which homicides were committed in two locations: Dade County (Miami), Florida, and Cuyahoga County (Cleveland), Ohio. The Dade County data included a total of 1,887 cases spanning the 15 year period 1956-1970. For Cuyahoga County 2,008 cases covering the period 1958-1970 made up the data sample. The time of injury to the nearest hour was known in most cases. The data were first subdivided by lunar phase and then combined according to phase in order to see if there were particular intervals during which more or fewer homicides were committed. The interesting results for the two counties are shown in Figure 13.1. Here the combined numbers of homicides are plotted according to the part of the lunar month in which they occurred. The homicide rate does not seem to have been uniform through the lunar month. For Dade county, two peaks, around full and new moon, are statistically significant. One of these is the 24-hour period beginning 24 hours after new moon (probability = .003); the other extends to one day before and after the full moon (p = .03). In the Cuyahoga County homicides there are four peak times of occurrence, none of which are as highly significant statistically as in the Dade County data. The two periods approaching significance (p = .07) are the 24 hours after new and the 24 hours starting 48 hours after full moon.

To lend further credence to this suggestion of lunar influence on homicide frequency, the authors have compared their Cuyahoga County results with another study showing that an apparently similar lunar periodicity shows up in the running activity of hamsters observed over a two-year period. The comparison given by Leiber and Sherin is shown here in Figure 13.2. The

184

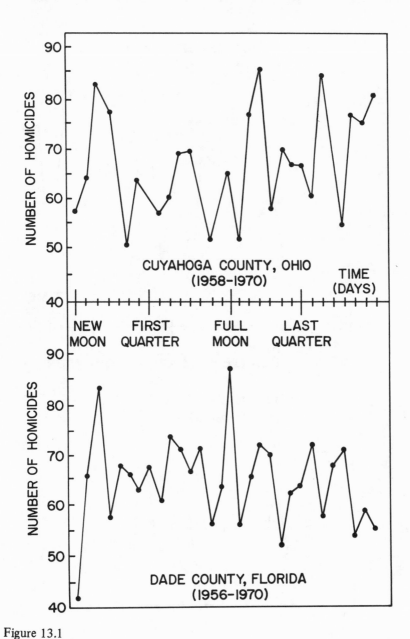

Figure 13.1

The number of homicides plotted versus lunar phase for (above) Cuyahoga County, Ohio (greater Cleveland) and (below) Dade County, Florida (greater Miami).

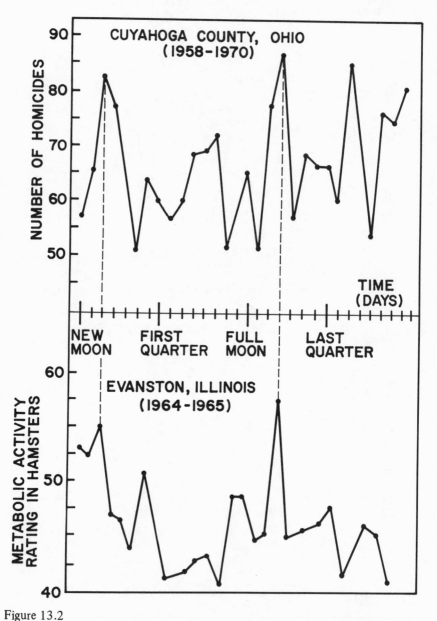

Figure 13.2

(Above) The number of homicides in Cuyahoga County plotted versus lunar phase and compared with a similar plot of the metabolic activity for hamsters in Evanston, Illinois, shown below.

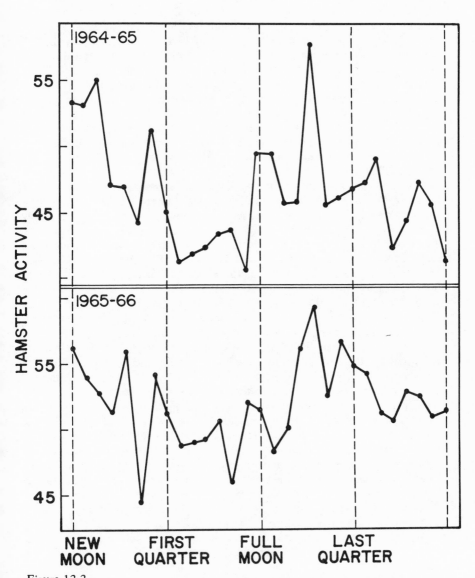

Figure 13.3

Plot of hamster activity in Evanston, Illinois versus lunar phase.

hamster study was done by Frank Brown and Young Park (13.17) who kept track of the amount of time spent on a running wheel by four male hamsters (one at a time) in 1964-1966. This was done in Evanston, Illinois, at a latitude very close to that of the location providing the Ohio homicide data so if there are some sort of gravitational effects involved here, you might expect the two locations to suffer the same lunar effects. But Brown and Park graphed the two years separately, Fig. 13.3, and Leiber and Sherin chose the best (1964-65) of the two cases for their comparison. Unfortunately, the corresponding curve for the other year (1965-66) does not show the new moon peak, and in fact the peak for the hamster curve that does agree is four days after full moon, and not three days as in Cuyahoga County. If the hamster data for the two years were to be averaged together the only significant variation would probably be the lunar month variation, one peak each around full and new phases. Brown and Park suggested that tides in the atmosphere, hence barometric pressure changes, or variations in the earth's magnetic field with lunar phase might be sensed by the hamsters. We mentioned above that barometric pressure changes in the middle latitudes are virtually undetectable, so not a likely agent, and in any case they most likely vary with the solar daily period as do their equatorial counter parts. The moon-induced atmospheric changes that are measureable also include a modulation of the height of the ionosphere, which in turn does produce small changes in the earth's magnetic field (13.18), so this is perhaps a more promising direction in which to look for an explanation. In any case the limited data for these few hamsters must be considered not very strong. The peaks of hamster activity do not match well with the homicide data. Then there is the non-trivial question of how running hamsters are related to homicidal humans.

But there are other difficulties which weaken the impact of the Leiber and Sherin study. For one thing the data are surely affected by the trend toward larger population. As a result there are higher numbers of homicides at the end of the 14 year period. For another they have not ruled out the possibility that other periodic effects may be spuriously producing the correlation they claim. You might expect, and there is ample supporting evidence, that more crimes of various sorts are committed on weekends. Many people have or want more money on weekends, have more time, drink more, etc. A check of the Charlottesville-Albemarle, Virginia rescue squad (13.19) reveals a far higher frequency of emergency calls on weekends than at any other time. The difficulty here is that in any year about one quarter of the full moons occur on Friday or Saturday. Any weekly periodicity will tend to become confused with or mistaken for a lunar correlation.

Alex Pokorny and Joseph Jachimczyk of Baylor College of Medicine attempted to replicate the Leiber and Sherin results through a study of 2,494 homicides that occurred over a 14-year period in Harris Country, Texas (Houston) (13.20). Although in part following a procedure much like that of Leiber and Sherin, they also looked at the homicide frequency by hour of the day and day of the week. Their results are plotted with respect to lunar phase and compared with the Leiber and Sherin study in Figure 13.4. Now the Dade and Cuyahoga County variations are correctly represented so the variations are not artificially exaggerated as they are as originally plotted by Leiber and Sherin.

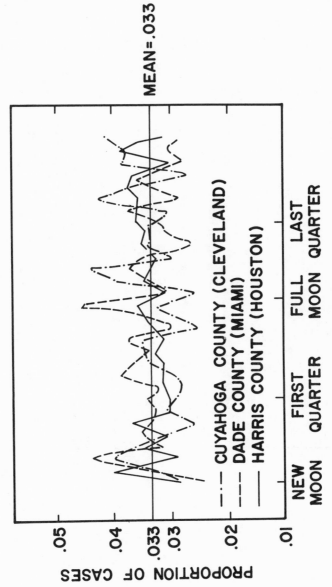

Figure 13.4   Simultaneous plots of homicide rates in the greater Cleveland, Miami, and Houston areas versus lunar phase.

189

TABLE 13.1   Number of Emergency Psychiatric Patients at Various Moon Phases, Janury 1, 1965 to December 31, 1966 at the Bronx Municipal Hospital Center (13.23).

	Full	Last	New	First
Total Number of Patients	595	609	604	588
Average Number of Patients per 24 hours	24.79	25.38	25.17	24.50

For Harris County there is no significant correlation with lunar phase. There is no correlation with Dade (similar latitude) or Cuyahoga Counties either. And the Harris County homicides are not statistically different over the lunar period from that expected purely by chance. They did identify a daily periodicity (most homicides between 11-12 p.m.) and a weekly variation (far more homicides on Friday and Saturday nights).

There has been one other recent search for a lunar synodic cycle in crime statistics (13.21). A total of 34,318 criminal offenses in nine categories (rape, robbery and assault, burglary, larceny and theft, auto theft, homicide, offense against family, drunkenness, and disorderly conduct) for Hamilton County (Cincinnati), Ohio in 1969 were divided roughly according to lunar phase. Full moon phase was considered to include the three days before and after the day of the full moon with the remaining days classified as "non-full moon" phase. Such a wide time window certainly will mask any possibly real lunar variations, especially in view of the dominant weekly cycle to be expected, or give a wrong impression of the true variations. While failing to quote actual numerical results, the authors claim "all relationships between the frequency of offenses committed and the phases of the moon, *except homicides* (italics ours) were significant (p < .01)." The crude analysis and short time interval of these data result in a study that adds no clarification regarding the issue of lunar influences on man's behavior. In fact no studies to date that we know of have used completely adequate analytic procedures in search of periodicities. The criminal acts-lunar phase correlations, at the very best, remain uncertain; on the face of it there seems no evidence of lunar effects.

We might finally ask if there is not an even more obvious place to look for evidence substantiating the folklore of lunar influences, that is in the rate of occurrence of psychiatric disorders. This is, after all, where the terms lunatic and lunacy arose. The 18th century English jurist Sir William Blackstone wrote the following definition:

> "A lunatic, or non compus mentis, is one who hath . . . lost the use of
> his reason and who hath lucid intervals, sometimes enjoying his senses
> and sometimes not, and that frequently depending upon the changes of
> the moon." (13.22)

Anecdotal reports of increased 'activity' at hospitals, on campuses, and elsewhere at the time of the full moon are easy to find. But the recollection of a busy night in a psychiatric ward around the time of full moon is not evidence — only the careful accounting of events can provide that.

The Bronx Muncipal Hospital Center operates a 24-hour emergency psychiatric counseling service serving about three-quarters of the Bronx. Every visit by a patient is registered and tabulated. In order to check for persons whose psychiatric disturbances might be correlated with the lunar cycle, Drs. Stephen Bauer and Edward Hornick (13.23) collected these data for a 24 hour period on the day of each new full and quarter phase of the moon over a two year period. If there were indeed more patients around full moon these numbers should show it. The results are given in Table 13.1. As you can see, and statistical tests confirm, no significant relationship appears between moon phase and the

191

number of patients visiting the emergency psychiatric service. They found only that there were fewer patients on Sunday. A very similar pattern was reported for a more limited study (13.24) at the Psychiatric Walk-in Clinic at Metropolitan Hospital, New York. After a three-month accumulation of patient statistics, there was no apparent lunar phase dependence. On the other hand a check of admission rate at the Ohio State University Department of Psychiatry in 1965 (13.25) showed somewhat mixed results. In this study the data were first combined by phase so that all admissions between first quarter and full, then full and third quarter, etc., were counted (Tabled 13.2a) over 12 'lunar months'. No significant differences were found. However, when the data were divided in such a way as to count the admissions from one mid-phase time to the next mid-phase, shifting the boundaries by about four days, it appears that a significantly greater number of admissions occur during the week of full moon (Table 13.2b). A breakdown of the data into six categories of disorder showed that two (psychotic disorders and transient situational personality disorders) were mostly responsible for the shifted peak. Here again the very broad time intervals (about seven days) and the small numbers weaken the case and do not rule out non-lunar related periodicities, especially weekly ones. There may be some fluctuation in the rate of admission, but the investigator may not have identified the correct cause.

In the most comprehensive study of this nature that we know of, Dr. Alex Pokorny (13.26) examined the times of occurrence of 4,937 psychiatric admissions for the three years 1959-1961. The data were divided into lunar quarter periods, quarter-to-quarter and midquarter-to-midquarter, in much the same way as the previously mentioned Ohio State study. An independent division according to the times of closest approach or greatest distance of the moon in its orbit was also made. The midquarter-to-midquarter data are tabulated in Table 13.3. These data for the hospital admissions show a significant relationship only for the year 1961 (p = 0.01). But it is especially interesting to note that each of the three separate years follows a different pattern. Thus, when the data are combined, so that any real effect should be enhanced, the variations smooth out leaving no trends of any consequence. The division according to lunar distance shows no pattern of variation. Noting that hospital admissions drop on weekends and that each lunar quarter is about seven days, Dr. Pokorny comments: "It therefore seems likely that the occasionally observed statistically significant relationship represents chance matching of these two weeklong cycles" (13.27).

Earlier in the chapter we mentioned the suggested correlation between lunar phase and weather as revealed in the pattern of rainfall. Since human behavior may possibly also be influenced by meteorological variables, we might ask if weather also can be related to deviant behavior. The literature attempting to relate suicide, homicide, and the weather is in fact extensive. One source estimates there are over 5000 technical articles published on the suicide-weather connection alone. Unfortunately the resulting picture remains unclear. Whereas some studies report a dependence of suicide and homicide on such factors as barometric pressure, humidity, precipitation, cloudiness, and especially season of the year (13.28), others have found none (13.29, 13.30). There have also

TABLE 13.2. Hospital Admissions at Various Moon Phase Periods for Columbus, Ohio in the interval December 29, 1964 to December 22, 1965. (13.25)

	New	First	Full	Third
a. Number of Patients in Interval	250	230	281	282
b. Number of Patients midpoint to midpoint	238	232	315	247

TABLE 13.3. Psychiatric Admission Frequency by Lunar Quarter from Midpoint between one phase to Midpoint of Following Phase (Houston, Texas). Expected numbers have been rounded to the nearest whole number. (13.26)

Phase	New	First	Full	Third
1959				
Admissions	385	372	345	392
Expected	356	352	385	401
1960				
Admissions	387	419	349	347
Expected	394	373	365	369
1961				
Admissions	429	453	565	494
Expected	473	484	585	479
1959-61 Totals				
Admissions	1201	1244	1259	1233
Expected	1225	1207	1252	1252

been related investigations of purely geophysical factors again some showing limited positive results (slightly higher psychiatric hospital admission rates on days of high cosmic ray activity, 13.31) and some showing negative results (fluctuations in the earth's magnetic field do not influence psychiatric hospital admissions, suicides, or homocides, 13.32).

In the search for factors which might influence human behavior all of the planets and the sun in concert have been involved in a study of some notoriety:

"In March 1951 John H. Nelson, an American electronics and radio engineer, published a sensational article . . . Nelson's article was an equally serious account of his research into factors affecting radio reception. But his report was to shatter orthodox views about humans and the Universe for his findings appeared to confirm the basic belief of astrology − that the planets can and do influence our lives" (13.33).

Nelson's paper "Shortwave Radio Propagation Correlation with Planetary Positions" (13.34) has been referred to frequently in the astrological literature. In his study he tried to show that radio communications over the North Atlantic were more likely to be disturbed during certain planetary configurations. Some astrologers have taken delight in pointing out Nelson's configurations are just the planetary aspects they themselves use. However, they gloss over the fact that Nelson chose to examine *heliocentric* planetary configurations instead of *geocentric*. Thus they are *not* the astrological configurations. Furthermore, it is not clear what radio propagation effects have in common with natal charts and astrological prediction in our lives. But these are minor points.

The physical connection of interest to us concerns the cycle of solar activity. It is well known that the numbers of sunspots vary with an approximately eleven year cycle. So do some other features of the sun, for example the outbursts known as solar flares. The ionosphere of the earth is affected by such solar outbursts. Since this layer of the earth's atmosphere reflects certain frequency radio waves, a consequence of its being disturbed is interference with radio transmissions. Nelson thought certain planetary relationships might have an effect on radio by somehow stimulating the sun.

Far from shattering "orthodox views" a careful reading of Nelson's article only reveals a very poorly executed piece of research. We wonder if the astrologers who so enthusiastically endorse the results ever actually read it. To bring up a few criticisms: there is virtually no discussion of the data used in the study; Nelson selected only four planetary aspects to examine ignoring all others where a better procedure would have been to take all radio disturbed days and see where the planets were; furthermore while counting the number of disturbed days corresponding to his selected "configuration days" he apparently ignored the configuration days when there were no disturbances, thus enhancing an uncertain correlation; and he did not consider the likelihood of the alleged correlations occuring purely by chance. Although the astrologers who allude to this work mention an 80% success rate for predicting radio conditions, in fact Nelson made this claim only for a simple procedure based on the direct telescopic observation and classification of sunspots and having nothing whatsoever to do with his planetary aspects.

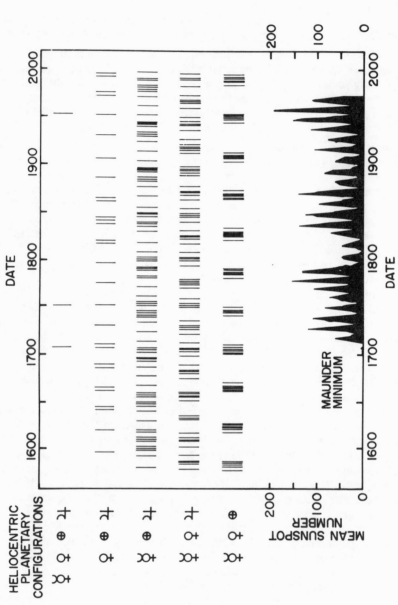

Figure 13.5   Heliocentric planetary configurations compared with mean sunspot numbers. Note the so-called Maunder Minimum in which virtually no sunspots were recorded for over 50 years in the late 17th and early 18th centuries

There have been others interested in connecting planetary configurations, tides in the sun raised by the planets, and the sunspot cycle (13.35). When two or more planets from certain geometrical arrangements, lining-up for instance, it has been argued the enhanced tidal pull on the sun somehow effects the number of sunspots. Nelson referred to such a mechanism. This all seems rather unlikely now thanks to a demonstration provided by the sun itself. Remarkably, for a period of about 70 years in the 17th and early 18th centuries there were almost no sunspots and apparently the 11-year cycle temporarily suspended operations. Still the planets moved regularly in their orbits sequentially aligning with one another and 'pulling' on the sun in exactly the same periodic fashion during the spotless interval as it does now. A graphical illustration of this is given in Figure 13.5 (13.36). Here the times of certain planetary conjunctions, the strongest tidal producing alignments, are marked along with mean sunspot numbers for the same period. The sunspots stopped while the tidal forces continued. To say the least, this casts considerable doubt on the tidal theory for the origin of sunspots and Nelson's claim for the indirect effects of planetary aspects on the earth as well.

Just to indicate in a broader context how complicated these questions regarding human behavior really are, a few comments regarding the better established physiological effects on living organisms arising from external factors might be of interest. To begin with, we know the number of air molecules with negative or positive electrical charge varies. This ion concentration is affected by solar radiation, cosmic rays, thunderstorms, and air pollution sources among other things. Exposure to significant concentrations of positive air ions can produce (through the release of the hormone serotonin) symptoms such as headache, tension, and reduced breathing capacity in many individuals (13.37). Daily, or circadian, rhythmic variations in the level of serotonin and a variety of other hormones such as adrenaline, insulin, progesterone, and testosterone are well established (13.37, 13.38). There are similar variations in body temperature, blood pressure, heart rate, and blood clotting time. Through its secretions, the pineal gland may be most important in governing or influencing these biological changes by controlling the activity of other endocrine organs, although the details are not yet clearly known. Other aspects of animal behavior are affected by these internal changes. In addition of the circadian rhythms, there are some indications of seasonal variations in the activity of the pineal gland, although variations in hormone levels have not been directly established. One moderator of such rhythmic behavior seems to be the amount of light in the environment due, for example, to the length of the day and its seasonal variation. A neural pathway linking the retina of the eye and the pineal gland by way of the hypothalamus (13.39) has been identified in a number of animals, thus providing a mechanism for this observed light dependence. Of further interest, the susceptability of organisms to electrical and magnetic fields is just beginning to be understood. In recent years it has become possible for the first time to measure the magnetic fields produced by the heart, brain, and lungs. They are about one billionth as strong as the earth's magnetic field (13.40). Exposure to relatively strong external magnetic fields has been shown to produce changes in the internal organs of mice similar to the effects of stress (13.41), to affect cell development and aging in plants (13.42), and to induce mutations changing the sex ratios of offspring in the fruit fly (13.43).

Human behavior is certainly the result of a complex interaction of many variables both physical and psychological. With so many factors influencing normal, as well as deviant, behavior, it is exceedlingly difficult to identify anything resembling a fundamental cause. This surely emphasizes a common difficulty in all of this. In evaluating apparent correlations, it should always be kept in mind that "a statistical relationship, however strong and however suggestive, can never *establish* a causal connection" (13.44). This is merely another instance of the fallacy of affirming the consequent (Chapter 2). Consider for example: if the moon affects the geomagnetic field, which is somehow related to the establishment of certain weather patterns, which in turn somehow affect the number of ions in the air, and subsequently the level of particular hormones in the body, which may cause some susceptible people to feel rotten, and perhaps even commit a homicide — can you blame the moon?

Although we can perhaps see a vague outline of a mechanism and causal chain of physical and biochemical effects on physiological processes and their resulting behavioral consequences, a great deal of painstaking study is still required. If any conclusions are to be drawn here regarding astrology, it is that the traditional systems should be discarded. The possible "celestial influences" we have discussed bear no resemblance whatever to the astrological predictions and descriptions. Biologists and chemists are doing quite well, thank you, and astrology has added nothing at all to our understanding of the behavior of living organisms. Once again, we find it is conventional science that has made progress in freeing us from the shackles of superstition.

13.1   Heaton, R.H., *Geophysical Jour. Royal Astr. Soc.*, **43**, 307 (1975).

13.2   Strahler, A.N., "The Earth Sciences", p. 92, Harper and Row Publishers, New York (1963).

13.3   Ibid., ref. 13.1.

13.4   Klein, F.W., *Geophysical Jour. Royal Astr. Soc.* **45**, 245 (1976).

13.5   Johnston, M.J.S. and Mauk, F.J., *Nature*, **239**, 266 (1972).

13.6   Mauk, F.J. and Johnston, M.J.S., *Jour. Geophys. Research*, **78**, 3356 (1973).

13.7   Bradley, D.A., Woodbury, W.A., and Brier, G.W., *Science*, **137**, 748, 1962.

13.8   Adderly, E.E., and Bowen, E.G., *Science*, **137**, 749 (1962).

13.9   Lethbridge, M.D. *Jour. Geophys. Research*, **75**, 5149 (1970).

13.10   Webb, H.M. and Brown, F.A., Jr., *Biol. Bull.*, **129**, 582 (1965).

13.11   Schone, H., *Gravity Receptors and Gravity Orientation in Crustacea*, p. 223, in "Gravity and the Organism", eds. Gordon, S.A., and Cohen., M.J., University of Chicago Press, Chicago (1971).

13.12   Gruner, L., *Ann. Zool. Ecol. Anim.*, 7, 399 (1975).

13.13   Idyll, C.P., *National Geographic*, 135, 714 (1969).

13.14   Robertson, D.R., *Comp. Biochem. Physiol. Comp. Physiol.*, 54, 225 (1976).

13.15   Danthanarayana, W., *Aust. J. Zool.*, 24, 65 (1976).

13.16   Lieber, A.L., and Sherin, C.R., *Amer. Jour. Psychiat.*, 129, 69 (1972).

13.17   Brown, F.A., Jr., and Park, Y.H., *Proc. Soc. Exper. Biol. and Med.*, 125, 712 (1967).

13.18   Chapman, S., *Geophysics*, p. 384, eds. DeWitt, C., Hieblot, J., and Lebeau, A., Gordon and Breach, New York (1962).

13.19   Slaughter, E., unpublished study.

13.20   Pokorny, A.P., and Jachimczyk, *J., Am. Jour. Psychiatry*, 131, 827 (1974).

13.21   Tasso, J., and Miller, E., *Jour. Psych.*, 93, 81 (1976).

13.22   Blackstone, W., quoted in Oliver, J.F., *Am. Jour. Psychiatry*, 99, 579 (1943).

13.23   Bauer, S.F., and Hornick, E.J., *Am. Jour. Psychiatry*, 125, 696 (1968).

13.24   Lilienfeld, D.M., *Am. Jour. Psychiatry*, 125, 1454 (1969).

13.25   Osborn, R.D., *Jour. Psychiatric Nurs.*, 6, 88 (1968).

13.26   Pokorny, A.D., *Jour. Psychiatric Nurs.*, 6, 325 (1968).

13.27   Ibid., Ref. 13.26, p. 327.

13.28   Eastwood, M.R., and Peacocke, J., *Brit. Jour. Psychiatry*, 129, 472 (1976).

13.29   Pokorny, A.D., Davis, F., and Harberson, W., *Am. Jour. Psychiatry*, 120, 327 (1963).

13.30   Porkorny, A.D., and Davis, F., *Am. Jour. Psychiatry*, 120, 806 (1964).

13.31   Friedman, H., Becker, R.O., and Bachman, C.H., *Nature*, 200, 626 (1963)

13.32   Pokorny, A.D., and Mefferd, R.B., *Jour. Nerv. and Mental Disease*, 143, 140 (1966).

13.33   King, T., "Love, Sex, and Astrology", p. 28, Barnes and Noble Books, New York (1975).

13.34 Nelson, J.H., *RCA Review*, March 1951, p. 26.

13.35 Wood, R.M., and Wood, K.D., *Nature*, 208, 129 (1965).

13.36 Smythe, C.M., and Eddy, J.A., *Nature*, 266, 434 (1977).

13.37 Sulman, F.G., "Health, Weather, and Climate", p. 50, S. Kanger Publ., Basel (Switzerland) (1976).

13.38 Reiter, R.J., *Psychoneuroendocrinology*, 1, 255, 1976.

13.39 Moore, R.Y., and Eiehler, V.B., *Psychoneuroendocrinology*, 1, 265, 1976.

13.40 Cohen, D., *Physics Today*, August 1975, p. 34.

13.41 Barnothy, M.F., and Sumegi, I., "Biological Effects of Magnetic Fields", p. 103, Vol. 2, Ed. M.F. Barnothy, Plenum Press, New York, (1969).

13.42 Dunlop, D.W., and Schmidt, B.L., Ibid., ref. 13.41, p. 147.

13.43 Tegenkamp, T.R., Ibid., ref. 13.41, p. 189.

13.44 Kendall, M.G., and Stuart, A., "The Advanced Theory of Statistics", p. 11, Vol. II, Haffner Publishing Co., New York (1972).

Chapter 14
Follies and Blasphemies

In the preceeding chapters, we have examined astrological experimentation, laws, theory, and prediction. In each instance we have found very little if any homage paid to the scientific method. Indeed, a rather significant number of astrologers seem to be quite capable of subverting that method's marvelous philosophical machinery. Although our discussion has almost exclusively centered on astronomy-astrology aspects, the astrologers' dealings with other branches of science have proceeded in a similar vein.

For example, in the concept of the triplicities the astrologers cling tightly to the Aristotelian view of four basic forms or "elements", i.e. earth, air, fire, and water, despite the fact that, as any elementary chemistry student can attest, each of these "basic elements" is itself made up of two or more of the 104 currently recognized chemical elements. The existence of each of these elements from hydrogen, the lightest and most abundant element in the universe, to the highly radioactive element Lawrencium, has been verified through countless chemical reactions and analyses. There is even the possibility that additional "islands of stability" exist for atomic nuclei which are even heavier than Lawrencium (14.1, 14.2). Each element of course can also exist in one of three basic phases, gas, liquid and solid, depending on its surrounding temperature and pressure. Thus, in the face of such overwhelming empirical and theoretical evidence, Aristotle's four element model holds little more than historical interest for the modern chemist. On the other hand, this same system still constitutes front-line symbolism for the modern astrologer:

"Certain elements combine while others are mutually destructive. Thus air increases the power of fire, but water destroys fire. Generally speaking, earth and water combine with reasonable amicability, as do fire and air. This is because water and earth are regarded as the lower elements and both of them pull downwards, whereas fire and air as higher elements both verge upward." (14.3)

Perhaps the most non-astronomical clash between science and astrology occurs in the life sciences, specifically in the area of genetics. Decades of experimentation have led life scientists to the result that all of the characteristics of a given life form are determined by cellular hereditary factors, called genes. The genes reside within the fine threadlike structures known as chromosomes, which in turn lie in the nucleus of every individual living cell. Without addressing the complexities of cellular and genetic biochemistry, the basic experimental result of interest here is that all of the inherited characteristics* of a human being are genetically imparted to that human being at the moment of *conception*. These characteristics are then shaped by that person's environment. These results stand in stark contrast to the basic claim of natal astrology that this impartation and shaping occurs at the moment of *birth*, some nine months later. The question of whether the astrological significance should be attached to the moment of conception or to the moment of birth is one that is not at all new to astrology. The issue has been around at least since the time of Ptloemy and was undoubtedly first raised many years before then. Despite this uncertain state of affairs, the astrologers down through the years have usually opted for the time of birth in casting their natal horoscopes. Once more, the decision has been made not on the basis of any viable experimental data, but rather as a bow to expediency. As anyone who has been involved in a paternity suit can attest, the exact moment of a child's conception can be a most difficult item to determine, while the time of birth is well-defined and easy to measure. Thus, for the astrologer who ideally requires times correct to the nearest minute for the "best" horoscopic results, the time of conception, while meshing well with the findings of the geneticists, nevertheless creates a distasteful set of problems which are neatly avoided by employing the time of birth instead.

Such behavior could be regarded by scientists with a certain amount of bemusement were it not for the fact that in claiming to be scientists, astrologers would foist upon us the ultimate astrological assault on our intelligence. And claim it they do. A quick trip down to the local bookstore reveals titles such as "Astrology: The Space Age *Science*" by Goodavage, "Astrology: The Divine *Science*" by Moore and Douglas, and *"Scientific* Astrology" by Manolesco.* Yet for centuries, like the barnyard animals in the story of "The Little Red Hen", the astrologers have in effect said "Not I" when it came time to freeze in a

---

*There is some evidence, by no means universally accepted, that even our behavior and personality traits can be explained within the framework of the genetic theory (14.4).

*Italics ours.

cramped space at a telescope on a cold night in an open observatory dome in order to gain astronomical data. They have said "Not I" when the time came to forge these and other scientific observations into meaningful laws and theories describing the physical world. They have said "Not I" when the time came to alter their laws and theories when predictions based on those laws and theories went awry. Now, after over three centuries of success the scientist is, by and large, held in very high esteem by society. A Harris Survey for January 1978, for example, showed that 91 percent of the public view scientists as having either "very great prestige" or "considerable prestige"* . At this point, the astrologers enter the scene eagerly saying, "I will, I will" in their desire to partake of the scientists' success and funding, despite the fact that as a group, astrologers have consistently been either unwilling or unable to embark on a scientific journey of their own. Indeed, the bulk of the astrological community does not even display a reasonable understanding of the scientific method, let alone put it to use in solving astrological problems. It is perhaps this basic deception above all else that has, in recent years, prompted the sort of scientific response which produces the Humanist statement and its 186 signatures.

It should be emphasized here that there is a relatively small handful of astrologers who are gravely concerned over the current status of affairs in astrological research. Dean and Mather, for example, in "Recent Advances in Natal Astrology" write

> "The current chaos in astrology is largely the result of a chronic infatuation with symbolism at the expense of reason. This is because the majority of astrologers reject a scientific approach in favour of symbolism (based on dubious tradition), institution, and holistic understanding." (14.5)

Unfortunately, the higher quality astrological works, such as "Recent Advances in Natal Astrology", are almost totally lost in the avalanche of astrological refuse which has been dumped on the public. In fact, the situation may well be such that to attain any measure of scientific credibility, the truly objective and serious astrologers may have to follow the centuries-old lead of the astronomers by placing as much philosophical distance as possible between themselves and what Russell (14.6) refers to as the "conjurers, thimbleriggers, and charlatans".

Kepler once wrote of astrology that:

> "No one should regard it as impossible that from the follies and blasphemies of astrologers there should emerge a sound and useful body of knowledge". (14.7)

Certainly the astrologers' oft-stated but oft-neglected quest for a truly better understanding of the relationship between human beings and the physical universe is a most admirable one, and, indeed, one that is shared by all of science itself. We have already seen, however, that the body of astrological knowledge, in its present form, is neither scientifically sound nor scientifically useful. If it is

* The lowest scoring occupations are salesmen and politicians.

not worthwhile scientifically, then does it necessarily follow that astrology should be branded as a negative influence in society? The astrologers, of course, would answer no. After all, goes the argument, who has the more positive role in society, the astrologer who deals caringly with a client on a one-to-one basis and is thereby able to assist that individual through a difficult period in his or her life, or the scientist who works on weapons development at a government laboratory. Such a question, of course, is yet another illustration of the astrological penchant for "proof" by individual examples, and as such is double edged. The scientist could just as well ask the same question regarding the scientist who develops a vaccine for a deadly disease and the astrologer who astrologically advises a client against a needed operation for a child, resulting in the child's death*. Clearly such case by case examples cannot of themselves decide the issue of astrology's net worth to society.

Often we hear the argument from astrologers that because of the durability and longevity of astrology down through the centuries, it "must have something going for it". The "something" is, of course, assumed to be of positive value, but in fact similar comments can be made for war, murder, slavery, and rape, all of which are also included in the list of humanity's eons-old endeavors. Given the overall status of contemporary astrology, one might even argue that the services provided by the world's third oldest profession are remarkably similar to those of the world's oldest. Such commentary, once more, however does not really settle the basic issue of astrology's value. To best make that ultimate judgement, it is necessary to pursue a different course.

The laws of modern science, because of the very method in which they are discovered and described, are deterministic in nature. The law of gravity does not incline, it *compels*. If we venture out onto the beach or ski slopes without adequate protection for our skin, we very quickly become acquainted with nature's laws of radiation in the form of a sunburn. We cannot control the degree of that burn exclusively by the use of our own free will. Indeed if such power were available to us, the suntan lotion industry would never have enjoyed its long-term success. This more or less fatalistic quality of the laws of nature creates the ultimate dilemma for the astrologer. If the astrologers' correspondences are to carry the same weight as the laws of the scientist, they must carry with them a dimension of compulsion. Despite the astrologers' pious protestations to the contrary, this sense of sidereal determinism permeates astrological thought. Sydney Omarr, for example, tells us proudly that "the wise man controls his destiny − Astrology points the way" (14.9). Yet it is also Sydney Omarr who tells us of the unsuccessful struggles of both Franklin Roosevelt and John Kennedy to break the Twenty Year Sequence. The implication is clearly made that there was virtually nothing that these men could do to stave off their alleged rendezvous with sidereal destiny. In a similar vein, West and Toonder recount two stories from astrologers Madeleine Montalban and Alan Leo in which the stellar sign-posts win the day in an unexpected

*Sir John Manolesco in fact describes just such a case in which the astrologer involved was taken to court over the matter (14.8).

fashion (14.10). Typical of such stories is the following example which was quoted in the local press by an astrologer who in turn attributed it to Ebertine:

"In Germany, a man once was told to be cautious of trains on a certain day. Believing his astrologer, he avoided the tracks all day, walking on the other side of town.

While pausing on the sidewalk, he was hit on the head with a small object and killed instantly.

The object: a toy train" (14.11)

Thus, just as surely as the toy train fell as a result of the law of gravity, our ill-fated commuter was felled by the "laws" of astrology.

The homage which astrologers pay to the concept of determinism is often subtle but nonetheless makes itself manifest in a variety of ways. Astrologer Al H. Morrison optimistically sees the day when astrological destiny will permit the young astrologers who are now busily mastering astrophysics and computer technology to "naturally" take over ultimate leadership in these areas (14.12).

In a more sinister example of astrological determinism, Martin Abramson reported several years ago that in the field of TV entertainment:

"When the so-called 'astrology addicts' get control of shows, they not only seek out performers deemed astrologically in tune, but discriminate against those who are not." (14.13)

"Astro-discrimination" was also one of the topics of discussion at the March 1978 National Astrological Association convention in Tucson. According to Gregg and Anna Howe of El Cerrito, California, who own a computer company which specializes in analyzing the charts of potential employees, many companies consult them about an applicant's horoscope *before* hiring that person for a job. Presumably such information thus plays an important role in the final decision regarding the hiring of the given individual (14.14).

The underlying concept, of course, implies that we are astrologically what we are and nothing can be done about it. It is particularly distasteful since it is merely a sidereal version of a refrain that has been used for ages to justify virtually every variety of discrimination practiced against human beings.

Perhaps the extreme example of astrological determinism came in 1576 when astrologer Girolamo Cardan felt obliged to commit suicide for the sake of fulfilling his prediction of the day of his own death!

Interestingly, this underlying fatalism places astrology on a direct collision course with the bulk of the religious sector, most of which holds that as human beings we are accountable for our actions on the face of the planet. The religious assault on astrology has thus proceeded with a characteristic intensity that seemingly can only be achieved when one fundamentally irrational system confronts another. The Reverend Kenneth Delano, for example writes:

". . . a belief in astrology is reprehensible for any good Christian. Whether the modern Christian realizes it or not, every dollar he or she

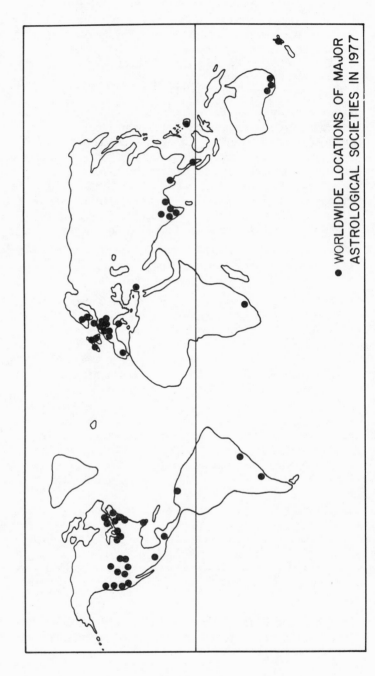

● WORLDWIDE LOCATIONS OF MAJOR
ASTROLOGICAL SOCIETIES IN 1977

Figure 14.1   Worldwide locations of major Astrological Societies in 1977, illustrating their almost total confinement to the non-communist world. The ultimate astrological synchronicity?

spends on a horoscope supports the cult of astrology and is a dollar paid in tribute to the ancient sidereal gods of the heathen nations . . . " (14.15)

In a similar fashion, Gary Wilburn warns us that:

". . . Astrology, therefore, is not eternal. One day the stars of heaven and their constellations will not flash forth their light; the sun will be dark when it rises, and the moon will not shed its light. When that happens what good will zodiacs and horoscopes do mankind? . . . the Bible says the astrologers will not even be able to save themselves much less their many followers." (14.16)

The extent to which human beings really exert a meaningful control over their own destinies is, of course, an age-old philosophical question. Certainly our freedom of choice is limited by our inherent ability. Not all of us, for example, are blessed with an operatic voice or a 90 mile-an-hour fastball, and the life choices available to the resident of a big city ghetto are demonstrably more limited than those available to a person living in middle class suburbia. So constraints on our free will exist, but should one's horoscope be included in the list of such factors? We have written this entire text detailing why we think not.

There is no doubt that astrology offers us a most attractive and convenient way out of life's responsibilities. Instead of assuming responsibility for our actions, we can simply say that "the stars weren't right" or even that "the stars made me do it". We suspect the reasons for the current return to astrology, as well as other occult systems, range from simple curiosity to a desperate groping for miracle solutions so the real problems of life and society may be avoided. Any such massive rejection of rationality stemming from ignorance of the facts, however, should be a matter of grave concern. A scan of human history reveals that when a society begins to embrace such irrational and fatalistic views, the end is close at hand. Thus Cicero's afore-mentioned skepticism of the Chaldean astrologers came at a time when Rome was vigorously and dynamically shaping her own destiny. Several centuries later, the Visigoths swept through a snivelling Roman Empire which had totally embraced that same astrology. The Fall of Rome has, of course, been blamed on everything from sexual debaucheries to the rise of Christianity, and with such marvelous historical factors lying about, it would be most presumptious of us indeed to set the blame for that Fall on the astrologers. Instead, we will borrow a page from the astrologers themselves and propose that the rise of astrology in a culture does not *cause* that culture's undoing, but rather is a sign or symptom of the conditions in a culture which betrays its inner weakness at that moment in history. So it was with classical Greece, imperial Rome, and medieval Christianity. Ironically, it is perhaps the ultimate astrological synchronicity of all, and, in light of the current astrological renaissance in the West, represents a most chilling correspondence indeed. There was once a time in the younger and more carefree days of human history when we could afford the luxury of an astrological dalliance. But now, faced with the awesome powers and problems of our technological adulthood, we can afford it no longer.

14.1   Nix, J., *Physics Today*, April, 1972, pp. 30-38.

14.2   Seaborg, G., *Ann. Rev. of Nuclear Sci.*, **10**, 53 (1968).

14.3   Hall, M., "Astrological Keywords", p. 43, Littlefield, Adams & Co., Totowa, New Jersey (1975).

14.4   Dawkins, R., "The Selfish Gene", Oxford University Press, Inc., New York (1976).

14.5   Dean, G., and Mather, A., "Recent Advances in Natal Astrology", p. 2, The Astrological Association, Bromley Kent, England (1977).

14.6   Russell, E., "Astrology and Prediction", p. 92, Batsford Press, London (1972).

14.7   Kepler, J., Tertius interveniens, cf. Gauguelin, M., "The Scientific Basis of Astrology", p. 157, Stein and Day Publishers, New York (1969).

14.8   Manolesco, J., "Scientific Astrology", p. 125, Pinnacle Books, New York (1973).

14.9   Browning, N., "Omarr: Astrology and the Man", p. 9, Signet Books, New York (1977).

14.10   West, J., and Toonder, J., "The Case for Astrology", p. 250, Penguin Books, Inc., Baltimore, Maryland (1973).

14.11   *The Fort Collins Triangle Review*, June 6, 1974, p. 2.

14.12   Morrison, A., *CAO Times*, Vol. 2, No. 2, p. 46 (1976).

14.13   Abramsom, M., "Have You Consulted Your Friendly Astrologer Lately?", *TV Guide*, October 4, 1969, p. 7.

14.14   Zodiac News Service, April 5, 1978.

14.15   Delano, K., "Astrology — Fact or Fiction", p. 115, Our Sunday Visitor, Inc., Huntington, Indiana (1973).

14.16   Wilburn, G., "The Fortune Sellers", pp. 53-54, G/L Publications, Glendale, California (1972).

Chapter 15

A Final Offer

During the course of this book, we have presented a more or less composite look at the affairs of contemporary astrology. The discussion has been far from complete, but given the many sides to astrology we could do little else in a text of this length. We are convinced however that astrology does not work. Astrology cannot be used to predict events of any kind, nor is astrology able to provide any useful information regarding personality, occupation, health, or any other human attribute. Having examined most of the available evidence, it seems to us these are the only possible conclusions. Nevertheless, recognizing the limitations inherent in the search for truth we admit there always remains the remote possibility of finding an astrologer or astrological technique that may succeed where all else has failed, especially since some astrological testing has not been based on a thorough synthesis of all factors in the horoscope. To allow for such an eventuality, we wish to offer the astrological community a chance to demonstrate they can do what they claim.

Such offers or challenges are not new. Author Philip Klass in his book "UFO's Explained" (15.1) has a standing contractual offer in which he agrees to pay $10,000 to any individual entering the contract upon the presentation of incontrovertible evidence that UFO's are extraterrestrial in origin. For each year that goes by in which no such evidence is forthcoming, the individual must pay Klass $100 up to a maximum of $1000. To date not one UFOlogist has collected on Klass' offer.

Similar opportunities have been presented to the astrologers. The June 3, 1973 issue of Parade Magazine offered professional astrologers a list of ten predictive challenges, many of which unfortunately, were not reasonable. One of the challenges, for example, was to "Name any famous person who will commit suicide and give the date within one week" (15.2). To not miss any possibilities in this case, the astrologer would have the gargantuan task of casting and

209

interpreting the horoscope of each and every "famous person" in the world, and even then there might not be such a suicide over the required time interval which extended only to January 1, 1974. Moreover each astrologer had to make predictions for all ten of the challenges in spite of the fact that they clearly covered various areas of astrological endeavor. In all fairness it is perhaps not too surprising that not even one astrologer responded (15.3).

Keeping the above factors in mind, our final offer to the astrologers is this:

Below are listed ten tests of astrological ability, all of which should be well within the framework of the claims made in the astrological literature. If an astrologer (whom we will define as a person whose income from astrology is 50 percent or more of his or her total income) wishes to accept one or more of these tests, the astrologer will notify Pachart House to that effect, and arrangements will be made to administer the test. Details of the experimental design will be set in advance to the mutual agreement of the parties involved. A deposit of 10% of the astrologer's fee will be required to cover the costs of the testing. For each test that the astrologer successfully passes, the authors and Pachart House will return the deposit and pay the astrologer's total fees for the casting of the horoscopes involved. This will be a fraction of the costs incurred, but if we are willing to donate our time to collect the basic data, the astrologer should likewise be willing to contribute his or her time to this experiment. On the other hand, for each test the astrologer fails, all charges for casting and interpreting the horoscopes will be forfeited, as well as the deposit. Thus, we will not stand to profit in any way from the astrologers' failures, but the successful astrologers stand to profit at our expense. The tests, then, are as follows:

1. Given the times, dates, and places of birth of sixty individuals, half of whom are violent criminals and the other half are peaceful, law-abiding citizens, identify at least 27 of the 30 violent criminals.
2. Given the times, dates, and locations of births of thirty individuals, identify the correct cause of death for at least 27 of the 30 from a list of thirty possible causes.
3. Given the times, dates, and locations of birth of sixty couples who have been married, but half of whom are now divorced, identify at least 27 of the 30 divorced couples. The time and date of the marriage will also be given to the astrologer in this case.
4. Given the times, dates, and locations of birth of sixty individuals, one half of whom have a given occupation, identify at least 27 of the 30 individuals in that occupation.
5. Given the times, dates, and locations of birth of thirty individuals and the personality profile for these individuals as measured by an acceptable standardized test (such as the Minnesota Multiphasic Personality Inventory), match each horoscope with a personality profile to a level significantly better than chance (p=0.001).

210

6. Given the times, dates, and locations of birth of thirty-two individuals competing in a single elimination tournament such as chess, handball, tennis, etc., determine the top four finishers in their correct order for the tournament. The nature of the competition, first round match-ups and dates and times of each of the tournament games or matches are also to be provided to the astrologer.
7. Given the times, dates, and locations of birth of sixty individuals, at least five of whom died in a common disaster, identify each of the victims and state the nature of the disaster.
8. Predict the date to within one day of the high point, low point, largest advance, and largest decline in the Dow-Jones Industrial Average over a one year interval of time.
9. Predict at least 30 days in advance the occurrence of an earthquake or any other natural disaster such as a flood, tornado, etc. in which the loss of life exceeds one hundred or more individuals, specifying the nature of the disaster, its exact time to within 24 hours and its location on the earth to within 100 miles.
10. Predict the existence and state the location to within one degree of an undiscovered body in the solar system having a diameter of at least 1000 kilometers.

Any attempts to pass these tests by non-astrological means will automatically constitute a failure. Any and all results will be publsihed by Pachart House at a future date.

In formulating these tests we have tried to purposely play to what the astrologers seem to claim as their strong suit abilities. West and Toonder, for example, tell us that:

"Given a chance, good astrologers can analyze, characterize and provide vocational and psychological guidance with impressive assurance."
(15.4)

If that is indeed true, then astrologers should have no trouble passing Tests 1, 3, 4, and 5. If the horoscope is the sensitive interplay of planetary influences that the astrologers allege, then there must also be an astrological Adams or Leverrier who can come forward and, using their wits and their pencils, show us the way to a new planet.

In short, we are offering astrologers the opportunity to horoscopically demonstrate the validity of their craft in a most convincing and unambiguous fashion. We don't think that they can.

211

15.1   Klass, Philip, "UFO's Explained", p. 422, Vintage Books, New York (1974).

15.2   Weisinger, M., "Is Astrology a $100 Million Hoax?", Parade Magazine, June 3, 1973, p. 7.

15.3   Mr. Pat Hammond, private communication from "Parade Magazine", 1975.

15.4   West, J., and Toonder, J., "The Case for Astrology", pp. 225-226, Penguin Books Inc., Baltimore Maryland (1973).

INDEX

213

SUN SIGN SUNSET

A Scientific Look at the Claims of
Sun Sign Astrology
By R.B. Culver

Perhaps the most familiar aspect of the current resurgence of astrology in our society is that of sun sign astrology. The basic premise of sun sign astrology is that an individual's overall characteristics, including personality, career, medical profile, and even physical appearance are at least generally related to the astrological sign of the zodiac in which the sun was located at the time of that individual's birth. It is the cornerstone idea behind virtually all astrological newspaper columns, astrological magazines, and indeed, most of contemporary astrology itself.

In "Sun Sign Sunset", Dr. Roger Culver of the Colorado State University, carefully examines this popular idea in terms of its history and development, internal contradictions, and most importantly, in terms of the various statistical studies which have been conducted to test the claims of sun sign astrology. In addition, Dr. Culver presents the results of statistical data which he himself has collected over the past twelve years, and concludes, with profound apologies to Sir Winston Churchill, that "Never before in human history has so much been based on so little by so many".

$ 9.95

Pachart Publishing House
1130 San Lucas Circle
Tucson, Arizona 85704